By The Editors Of Consumer Guide®

Getting Pests To Bug Off

By Stanley Rachesky

Contents

Manufactured in the United States of America
1 2 3 4 5 6 7 8 9 10

Published by
Crown Publishers, Inc.
One Park Avenue
New York, N.Y. 10016

Library of Congress Catalog Card Number: 78-538-90

ISBN 0-517-532697

Cover Design and Chapter Opener Art: Zimnicki Design
Pest Illustrations: Steven Boswick

About the Author

Stanley Rachesky is one of the nation's leading authorities on insects and pesticides. Perhaps best known for his weekly column "Pest-a-Side" in the *Chicago Tribune,* he is a popular lecturer and has contributed many research papers to the scientific literature. A former college instructor, Rachesky is now an urban and industrial entomologist for the University of Illinois Cooperative Extension Service. In this capacity he works with many urban groups, providing advice on urban and suburban pest control.

The Astonishing Insects

If insects didn't have so many natural enemies, they would rule the world. They are able to live almost anywhere. Vast numbers and programmed behavior are the keys to their success.

ALL FORMS of life on this planet belong to one of two categories—the plant or the animal kingdoms. In general, distinguishing a plant from an animal is quite easy. A plant is normally permanently attached to one spot, feeds (with some exceptions) on inorganic foods, contains chlorophyll, and has cell walls made of cellulose. An animal, on the other hand, is usually free-moving, feeds on organic foods, contains no chlorophyll, and has cell walls made of protein. There are some minute life forms that exhibit both plant and animal characteristics, making it impossible to draw a definite dividing line between the two kingdoms.

More than one million animal species have been discovered and named. About 75% of these known species are insects. This enormous variety of species enables some form of insect to live almost everywhere on earth. In addition, their tremendous ability to reproduce guarantees survival despite high death rates. It has been said that the number of insects is so vast that the combined weight of all the insects is greater than the combined weights of all other animals—including man!

Another factor in their success is the programmed behavior patterns that govern insect life. Because they lack reason and judgment, and thus are ruled by instinct, they are incapable of developing a civilization, but they are equally incapable of recognizing defeat. *Despite all the technological advances of modern man and all the insecticides now in use, not a single insect species has ever been exterminated.*

Development

All insects develop from eggs. A single female can lay up to one million eggs in her lifetime. The offspring usually hatch after the eggs have been laid. There are some species, however, that hatch internally. In general, eggs are deposited in places where the young have the best chance of obtaining food. The egg stage lasts between eight hours and two weeks, but the eggs of many species are viable for months or even years.

Once the egg hatches, the insect grows in different stages. There are four general categories of development:

Complete Metamorphosis—Metamorphosis is a complete change in form between birth and maturity. Most insects undergo this process. The insect develops from egg, to larva, to pupa (cocoon stage), to adult. The larva usually looks quite different from the adult, making it difficult for the average person to realize that these two dissimilar forms are actually the same insect in different stages.

During the larval stage, the insect has a voracious appetite and develops quickly. The pupal stage, in contrast, is a quiet time. The larva retires to a hidden and protected spot while complex chemical changes take place. Sexual organs, mouthparts, compound eyes, wings, and antennae develop, and the adult insect emerges, ready to carry on its tasks of mating and reproducing. Once the new generation has been laid, the adult usually dies. Common insects in this category are the June bug and the May beetle.

Incomplete Metamorphosis—The young of certain species grow in water and are known as naiads (a Greek word originally referring to water nymphs). Their bodies are adapted to surviving in an aquatic habitat. They have gills for breathing, climbing legs, and bodies streamlined for swimming. The adults of these species are terrestrial, are strong fliers, and have different feeding habits. An excellent example of this category is the dragonfly.

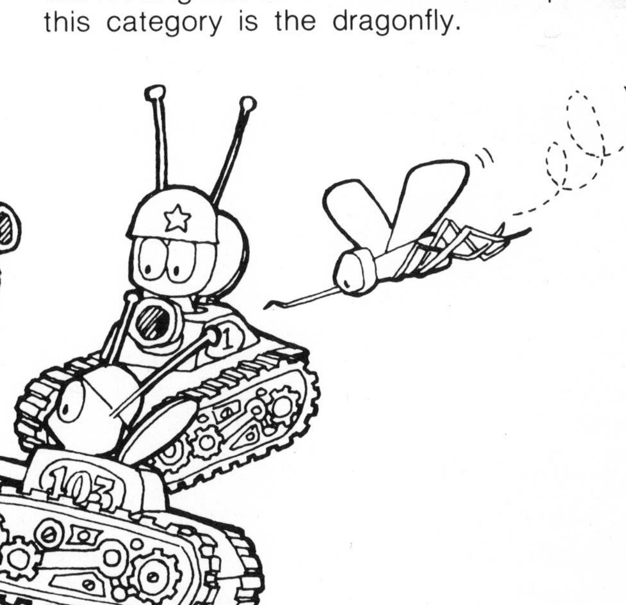

Gradual Metamorphosis—This type of insect undergoes a less radical change. The young usually have the same feeding habits as the adults. Commonly encountered insects in this group include grasshoppers, cockroaches, crickets, termites, and aphids.

Without Metamorphosis—Some insects never really change between hatching and adulthood. The young insects are readily recognizable as immature forms of the adult. Many common problem-presenting insects such as silverfish and body lice fall into this category.

Anatomy

The body of an insect is more or less an elongated tube with appendages such as legs, wings, or antennae, arising from a hard outer skeleton (exoskeleton). It is bilaterally symmetrical. What's on one side is on the other. It is segmented or jointed into three distinct body regions—head, thorax, and abdomen.

The Head

The typical insect's head is made up of a number of segments solidly fused together to form a hollow capsule. The insect's sense organs are located here. The antennae are used for smell, touch, and hearing. They are almost always segmented and can be used to identify the insect. The insect's eyes are also located on the head. Insects cannot move or focus their eyes. The eyes are made up of many small lenses or facets. There can be as many as 50,000 facets per compound eye. The light penetrating the insect's eye must come in perpendicularly to reach light-sensitive areas. What the insect actually sees are changes in shadows which alert it to danger. Some insect eyes are color-sensitive.

The insect's mouthparts are also located on its head and are often used in identification. The type of mouthparts reveals the feeding habits of the insects. There are six main types of mouthparts:

Chewing—These are considered the most primitive and basic variety and are found on such insects as cockroaches, termites, silverfish, beetles, grasshoppers, crickets, biting lice, and the larvae of moths and butterflies.

Rasping-Sucking—Only one insect—the thrips—has this type of mouth. The thrips lacerates or rasps the epidermis (surface) of plants with three needle-like organs known as stylets and then sucks the sap from each individual plant cell. The damage caused by thrips is called "silvertop" because the leaf of the plant takes on a silvery appearance.

Piercing-Sucking—Insects such as leafhoppers, aphids, and some species of flies have the ability to pierce cell walls and suck the blood of animals or the cell fluid of plants.

Sponging—Mouthparts of this type are found on insects that cannot bite. They are usually unable to pierce the skin of animals or the epidermis of plants. They must feed on liquid materials. Houseflies, blowflies, and fruitflies are good examples.

Siphoning—These mouthparts are found on adult moths and butterflies. They have a tube or tongue that sucks up liquids and remains in a coiled position when not in use.

Chewing-Lapping—Bees and wasps have mouthparts of this type. They can suck up nectar, model wax, and chew solids.

The Thorax

The middle section of the insect's body is known as the thorax. It's a very hard, tough region where many muscles are attached. They operate the three pairs of legs and the wings. Slit-like openings called spiracles are also located here. These are the external openings to the insect's respiratory system. The thorax is connected to the head by the cervix.

The leg of an insect is always jointed and consists of six parts: coxa, trochanter, femur, tibia, tarsus, and pretarsus. Insect legs are modified for running, jumping, grasping, digging, carrying, or swimming.

The wings of insects are unique in the animal world. They are an outgrowth of the body wall on the last two thoracic segments. They are the only invertebrates (animals without backbones) to have wings and the only animal in the world to have two pairs. Wing size and vein pattern and the position in which the wings are held are all helpful characteristics used to identify insects.

The Abdomen

The abdomen is the third and final major region of the body. It is made up of 11 segments, 10 of which are easily distinguished, while the 11th is greatly reduced. Spiracles (openings to the respiratory system) can be found on segments 1 through 7 or 8. The reproductive appendages are located on the last abdominal segment.

Internal Structure

It's really not necessary for the average layman to be concerned about the internal workings of the insect's organs. What is important is that an insect is an animal and has certain basic biological functions to maintain in order to carry out its life. Insects have a respiratory system, and although it does not work like ours, it still allows the exchange of oxygen and carbon dioxide to take place through the spiracles located on the thorax and abdominal segments. The movement of air through the spiracles can be compared to air movement by a fireplace bellows. During normal insect movement, air is exchanged.

The circulatory system of the insect is known as an open system because the insect's blood is not contained in veins or arteries. The insect heart consists of one dorsal vessel or a series of chambers. The heart pumps the blood and pours it over the insect's brain.

The digestive system works very much like ours. Food is taken into the mouth and moved down into the esophagus (crop), to the gizzard. The gizzard has a grinding and straining function for hard substances. The crop serves as a food reservoir. The food then moves toward the intestine, colon, and rectum where the waste material is excreted.

The excretory system takes waste materials and dumps them into the hind intestine. These waste materials are then excreted.

The nervous systems of insects vary greatly. In general, the nervous system resembles a piece of string with a series of knots tied in it. The string represents the spinal chord while the knots represent ganglia. Each ganglion is like a separate brain. In the higher insects, the ganglia merge to form concentrations of nervous tissue.

The reproductive system in insects is also similar to that of higher animals. The males produce sperm in their testes. The sperm travel through the vas deferens to a sperm receptacle where they may be stored. Upon copulation, the sperm are ejaculated into the female's vaginal opening, called the seminal receptacle or spermatheca. The female produces the egg which is fertilized by the sperm. The external genitalia, particularly those of male insects, are used to identify many species because they exhibit more distinct characteristics than any other structures on the insect's body.

People Pests

These are the flying, scurrying, wriggling, and crawling creatures that drive you crazy. Whether you live in an apartment, a house, or a mobile home, you are bound to encounter some of these pests. Nature-lovers often have trouble with outdoor insect nuisances as well. You don't have to share your home with these tiny trespassers — most can be easily evicted.

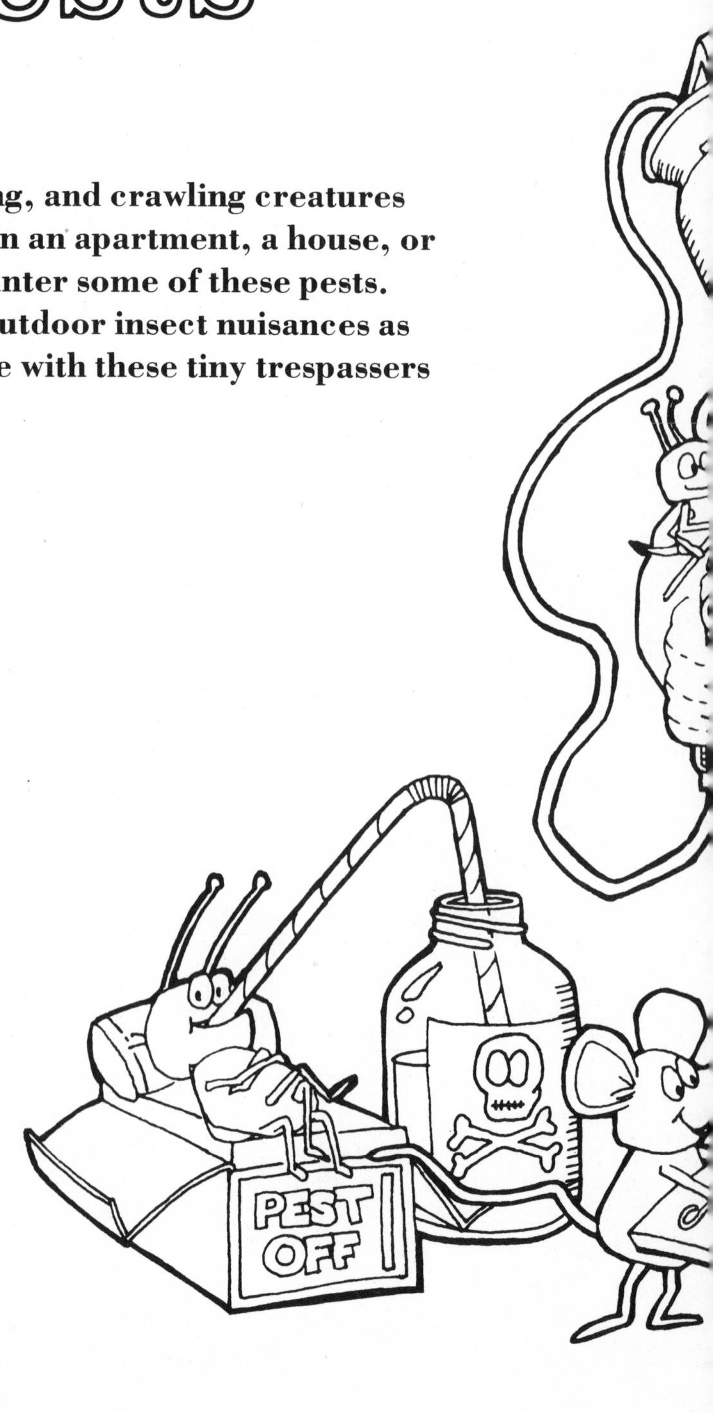

JUST THE thought of insects is enough to give most people the creeps. The situation is much worse, however, when you actually find them crawling around the house, eating your food or clothes, or inflicting painful bites.

Whether you have been invaded by relatively harmless creatures like silverfish, disgusting ones like cockroaches, or potentially dangerous ones like bees, it is possible to get rid of them. This chapter describes the pests that bother people and their possessions. A number of insects that live outdoors have been included because they can be real nuisances when you're picnicking, camping, or just enjoying the backyard.

Entomophobia

Entomophobia is the scientific name for fear of insects. Although most people find

them distressing, a few people actually live in terror. Their lives are made miserable by bugs that don't really exist. They isolate themselves to avoid contamination and spend a lot of time cleaning and spraying spotless and bug-free houses. They also suffer from itching and bites that have no basis in reality.

Do You Need a Professional?

If you don't want to handle the situation yourself, a professional exterminator can do it for you. CONSUMER GUIDE® recommends calling in a professional for termites. Also, people with known or suspected allergic reactions to bee and/or wasp stings should get help when these insects become a problem. When selecting an exterminator, it pays to comparison shop (ask for an estimate and about the method of control used). Make sure the company belongs to the state or national pest control association. This is usually a good indication of reliability.

In most cases, however, you can be your own exterminator. Simply identify the pest, read about its habits, and follow the control suggestions in the chart at the end of the chapter.

If you live in an apartment or condominium, you may have some special difficulties. Pests are seldom limited to one apartment and any control measures you take will not be effective against the bugs next door. Sooner or later they will return to reinfest your home. In order to eliminate these pests, the entire building must be treated.

Your first step is to check your lease and the local ordinances dealing with such contracts. In many areas, pest control is the responsibility of the landlord, not the tenant. Even if this responsibility is not clearly defined, a tenants' committee may persuade him to take control measures anyway. An agreement by all the occupants to coördinate their individual control efforts may be effective if the landlord takes no action.

Ants

Ants are among man's most common insect visitors. Their nests can be found almost anywhere—in your lawn or garden, in the soil next to the house, even behind a stove or refrigerator. About 7600 living ant species have been described, but it is believed that several thousand other species remain to be identified. Their habitats range from the tropics to the Arctic, and their feeding habits are equally varied.

Ants enjoy the good life. You already know how much they enjoy picnics, but they're just as happy with a piece of candy in the kitchen. Some species even maintain biological sugar factories; they transport aphids from plant to plant and collect the sweet "honeydew" secreted by the aphids in return.

Don't expect ants to pay attention to property rights—your home is their private estate in their opinion. They let you know it by building their nests wherever they find a convenient spot. These social insects, however, do have some redeeming qualities. Even though they eat your vegetables and flowers, they also feed on scales, mealybugs, and other plant-sucking insects. In addition, outdoor ants (which most people find much more tolerable than indoor ants) are fascinating to observe.

Ants have a complex social structure and a "language" with which they communicate. There are three basic castes: winged, fertile females (called queens); wingless, infertile females (workers); and winged males. Some species have other specialized castes as well. A few species even use other insects as "slaves." Although most insects have short life spans, a queen may live as long as fifteen years, laying generation after generation of eggs.

Carpenter Ants

If you have keen ears, you may hear a faint clicking inside your walls in the spring. You are probably eavesdropping on the conversation of two or more carpenter ants. These large black insects (Color Figure 29) communicate by clicking their jaws together. Despite their fondness for talking, they prefer chewing.

The carpenter ants' mission in life is hollowing out large "galleries" in wooden beams, joists, and floorboards, thus providing a comfortable amount of living space for themselves. They're especially fond of damp wood, so their presence may indicate that something has sprung a leak. Check your pipes, roof, and window sills.

Controlling these creatures can be tricky. Locating their nests is often a problem, because they tend to build inside the walls, but direct treatment of the nesting area is vital. One clue is a little pile of sawdust, much like that found inside pencil sharpeners. If you're lucky they may be in plain sight, but they are more likely to be hidden between wall studs or high on a foundation ledge in the basement. You may have to drill a hole near the nest.

Dusts are preferable to sprays because they provide better coverage. Spraying the ants' favorite paths, once you find them, is helpful. Complete control often requires several treatments, so be patient.

Carpenter Ant

Common Ants

Common ants, also known as pavement ants and sweet ants, are the little guys you'll find crawling over the peanut butter and jelly sandwich the kids dropped under the couch. They spend most of their time foraging for food to take back to their subterranean nests.

Controlling these ants is usually fairly easy. Since their nests are close to the foundation of your home, a foundation spray generally provides effective and safe control. A house of average size will probably require a minimum of four gallons of diluted spray, about one gallon per side. Spray along the foundation, covering the wall up to a height of four inches and a two- to four-inch strip of soil. Crawl spaces may also need spraying. Make sure you achieve thorough coverage. Spray all the way around the building, paying special attention to areas behind steps and any cracks that may provide entry.

Pharaoh Ants

The pharaoh ant is a prime example of the adaptability of the insect world. This species survives by living snugly inside building walls. They seldom live in large colonies; little groups are scattered throughout a building in various crevices and wall cavities.

Because there is no central nest, eliminating these creatures requires some special measures. At a large Midwestern zoo, for example, the pharaoh ant population became so large that it posed a serious threat to the animals living in the reptile house. Zoo officials suspected that the ants were actually feeding on baby reptiles.

Getting rid of the ants without harming the reptiles presented some problems. It was necessary to trick the ants into destroying themselves. Regular feeding stations for the ants were established. Once they became accustomed to foraging at the stations, poisoned ant baits were substituted for the food. These baits are effective because the ants bite off portions of them and carry them back to their nests.

You can use the same technique in your home. We do not recommend it, however, if you have small children or pets who might be harmed by the baits.

Aquatic Insects

These bothersome creatures, also known as swimming pool pests, can really spoil your outdoor fun. Some insects such as moths, leafhoppers, and beetles occasionally fall into a pool by accident. They're probably even more upset by it than you are! Other insects, however, are aquatic by nature—during part of their life cycles at least—and seek out ponds, streams, and pools. Mosquitoes and midges, for example, may breed in pools that are not properly maintained. The slime that builds up on the sides and bottom of pools is very attractive when egg-laying time approaches. Other insects you may notice in or around the pool are the backswimmer, the giant water bug, and the water boatman.

Backswimmer—The backswimmer is usually found floating on the surface of the water. You can identify it by the fact that it swims upside down with its long oar-like legs stretched out for a quick getaway. It feeds on other insects and an occasional tadpole or tiny fish. Try to avoid it because a backswimmer bite feels like a bee sting.

Giant Water Bug—This large (about two inches long) brown insect has the shape of a flattened oval. It feeds on insects, snails, tadpoles, and small fish. Because it is strongly attracted to light, it is also known as an "electric light bug." Whatever name you call it by, its bite is painful.

Water Boatman—This insect, which closely resembles the backswimmer, feeds on algae and microscopic creatures. It is about one-fourth inch long and dark gray in color. Its two front legs are flattened into scoops, its middle legs are long and slender, and its hind legs are flattened and shaped for swimming. The water boatman does not bite people, so it is more unsightly than annoying.

Control

If you find any of these creatures in your pool, a dip net will allow you to remove them without being bitten. Never use an insecticide in the pool itself. Proper filtration, cleaning, and chlorine treatments will keep the pool fairly insect-free. Drain the pool or arrange for maintenance if you plan to be away for two weeks or more. Children's wading pools should be emptied after each use. Lights often attract insects, so use them sparingly if safety considerations permit. If lights are necessary, yellow bulbs, set back from the pool, are less likely to attract pests. Shrubbery and grass near the pool should be kept closely trimmed. These areas may be lightly sprayed if insects are becoming a nuisance.

Assassin Bugs

These unpleasantly named insects have distinctive beaks that they can fold under their heads when they aren't feeding. A few species inflict painful bites. Among these is the easily recognized wheel bug, which has a peculiar spoke-like structure on its back. Assassin bugs are also known as "kissing bugs," a charming name derived from their decidedly uncharming hab-

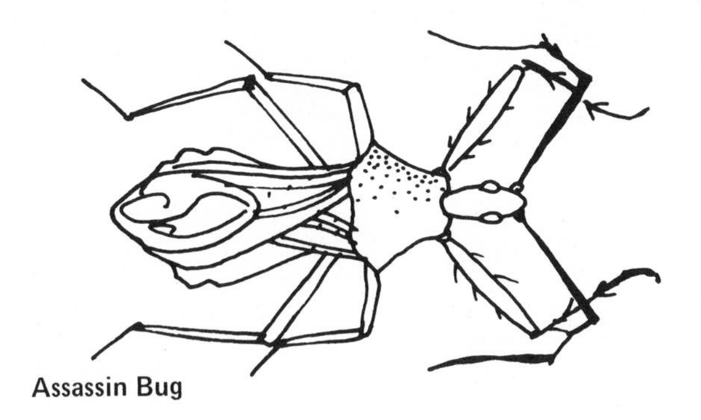

Assassin Bug

it of biting sleeping people near the mouth.

Assassin bugs are a serious problem in Latin America where they play an important role in the spread of Chagas' disease, an often fatal tropical illness characterized by prolonged high fever, edema (retention of water in connective tissues), and enlargement of the liver, spleen, and lymph nodes. The wastes of the bugs become infected with disease organisms and enter human bodies through open wounds. They may enter through the bite itself, cuts or scratches, or penetrate eye and mouth membranes.

The presence of assassin bugs may mean that rodents are nesting in or near your home. Control measures should begin with the removal of rat, ground squirrel, and other rodent nests. If necessary, spray baseboards, crevices, and thresholds inside the house and along the outside foundation.

Batbugs

Batbugs are often confused with bedbugs because they are found in mattresses and bedding. You are unlikely to be bothered by them unless you have suffered an infestation of the bats on which they live. Getting rid of the bats is obviously the first step to take (see "Small Animal Pests" for some suggestions). If the batbugs have invaded your living quarters, spray bedframes and springs thoroughly and lightly dust the seams and tufts of the mattresses.

Bedbugs

Bedbugs have been keeping company with man throughout history, along with lice, cockroaches, and rodents. These reddish-brown creatures are also known as chinches, red coats, mahogany flats, and wall lice (Color Figure 12).

They are nocturnal feeders, leaving their homes in search of food as evening approaches. They live along baseboards, under wallpaper, or in beds. Bedframes, mattresses, and springs are among their favorite hiding places. Bedbugs often migrate from one apartment to another through the wall openings where pipes run. Most bedbugs enter a building along with the secondhand furniture. If you purchase any used furniture likely to harbor these creatures, be sure to treat it before bringing it into your home.

Control measures should be taken at the first sign of any problem. Pay special attention to mattresses (checking all seams and buttons carefully), bedframes, upholstered furniture, behind pictures and wall hangings, and wall and floor crevices. One thorough treatment should provide six to eight months of complete control.

Bedbug

Bees

Bees, despite their great usefulness, tend to be unpopular—probably because they are so well-equipped to defend themselves. They are, however, among the most interesting creatures. Not only are they responsible for pollinating a great variety of plants and the creation of honey, but their complex social structure is fascinating. Even the great Sherlock Holmes, you will remember, spent his later years keeping bees on the Sussex Downs.

Bumblebees

Bumblebees may present a problem when they nest in or near buildings. These large, hairy insects like to reside in old mattresses, car cushions, and similar materials.

Carpenter Bees

Unlike most species which are very social, carpenter bees are loners. They rarely sting, but their large size makes them frightening to encounter. You may find their nest holes in soft woods around your house.

Honeybees

Honeybees have been considered valuable for thousands of years. Long before sugar was common, people satisfied their taste for sweets with honey.

You may see honeybees swarming in the spring or early summer when new colonies are formed. Colonies are composed of three castes—a queen, workers (infertile females), and drones (fertile males). When a colony divides, the old queen establishes a new hive while the stay-at-homes wait for a new queen to hatch. The swarming honeybees are unlikely to bother you if you keep your distance.

The situation becomes more complicated when the swarm decides to set up housekeeping inside a building. Once the honeybees settle in, they begin building combs. It is best to remove them before this process gets underway.

Beehive Between Wall Studs

Control

There is always an element of risk in dealing with bees. If you have ever seen a picture of a beekeeper at work, you probably remember the heavy veil and gloves he wore to avoid stings. Perhaps the bees have a right to be furious—after all, you would be upset if someone broke into your home to steal the groceries. If the thought of dozens or even hundreds of bees makes you nervous or if you have a history of allergies, consult a professional. A local beekeeper may want the colony. If beekeepers are scarce in your vicinity, an exterminator can take care of the problem.

Insecticides are the safest means of control, but you must know the exact location of the honeycomb before using them. If the nest is not close to the flight entrance, the insecticide will not reach it. Tap the wall at night and decide where the buzzing is loudest. Since the interior of the nest is usually about 95 F, you may be able to feel the heat through the wall. You can doublecheck by drilling a small hole into the suspect area. If the drill bit has honey on it, you've found the nest (Color Figure 26).

You can use either sprays or dusts, but as a rule, dusts are preferable. Apply the insecticide at night—through the entrance hole if the colony is close to it or through a hole drilled in the wall above the colony. Next, seal the exits so that the bees cannot escape. A very large colony may need a second treatment ten days later to kill newly emerged young bees.

Remember that you will have to open the wall and remove the dead bees, comb, and honey. If the nest is not removed, it will attract carpet beetles, cockroaches and ants, and the decaying honey may seep through the wall or ceiling. Wait until all sounds of activity have stopped before attempting to remove the nest. Wait two or three days to be on the safe side. Once you have the nest, dispose of it by burning or burial. Because it is toxic, it should not be thrown out with the trash.

Bee Stings

If you encounter a bee or two, there is no need to panic. They are much more interested in pollen than in you. Indoors, an aerosol spray will kill it quickly. If one flies into your car, you can gently herd it out with a map or newspaper, or crush it with a glove, handkerchief, or rolled paper.

A bee sting is painful but seldom dangerous. Most people experience some swelling. A sting near the eyes or mouth will produce a noticeable local reaction that usually subsides in a few days. An icepack or local anesthetic will reduce discomfort. About 2% of the population, however, may have far more severe reactions. For those people, untreated stings may be fatal. Such individuals should always have medication close to hand. Anyone who has been stung in the past should be very careful, because hypersensitivity may develop after the first bite.

Bluebottle and Greenbottle Flies

These pests are just as annoying as houseflies. You can identify them by their metallic blue or green abdomens. They breed in the decaying bodies of dead animals and have been known to travel more than 25 miles from their breeding grounds. These flies are particularly abundant in areas with inadequate sanitation, so keep pet droppings cleaned up and drain any standing water near your home. Spray shrubbery and tall grass if necessary. Impregnated resin strips fastened to the lids of your garbage cans will eliminate breeding.

Booklice

Booklice are literary bugs that make their home on books, especially those stored in basements or other damp areas. When their population booms, you may spot them crawling over furniture and food. No matter how plentiful they are, however, damage is usually negligible because they feed primarily on microscopic molds. They will not harm people. They also live outdoors in grass and leaves, under bark (which is why they are sometimes called barklice), or in other moldy environments.

These little (about 1/16-inch long) pests can be difficult to eliminate. The problem is usually outside the house. You can discourage them from entering by keeping books and papers out of damp areas. If you have any mold growing on basement walls, improving ventilation and lowering the heat may help. Spray shelves or other areas where you see booklice.

Boxelder Bug

Boxelder Bugs

The boxelder bug (Color Figure 40) becomes an indoor pest in the fall. During the summer months, these little creatures spend their time outdoors happily feeding on boxelder seeds. You can also find them on maple, ash, and fruit trees and even on grape vines.

As cool weather approaches, they migrate into buildings for warmth and protection, clustering on the sides of houses and crawling into any available crevice. On warm, sunny midwinter days, they like to sun themselves on the south and west sides of the house. Once they get into the walls, they hibernate. Some, however, move right inside and start to bother the residents.

Although they don't attack people or clothes, they do feed on some houseplants. Perhaps their most annoying habit is soiling curtains and wallpaper with fecal deposits.

Insecticides are the most effective means of control. Spray tree trunks and the surrounding ground in the fall if you see these bugs. In addition, spray the house foundation and a three-foot strip of ground around the sides.

Carpet Beetles

The black carpet beetle is one of the most destructive of all indoor pests. The adult does no damage, preferring flowers to fabrics, but the larva can wreck your winter wardrobe (Color Figures 6 and 7). Its favorite food is wool and it loves to gorge on your pants, sweaters, and brand-new coat. Even wool-synthetic blends provide a tasty snack.

Adult Carpet Beetle

Recent tests using nine fabric samples revealed that a blend of 50% wool, 20% polyester, 19% nylon, and 11% cotton was irresistible. When separate wool and synthetic yarns are woven together, the larva eats the wool and ignores the synthetic fiber. When the fabric is woven from a yarn made of blended wool and synthetic fibers, however, the larva swallowed both fibers, later excreting the undigested synthetic fibers. Any blended fabrics you purchase, therefore, should be treated as carefully as all-wool fabrics.

Washing or dry-cleaning eliminates this pest. Pay special attention to pockets, cuffs, and seams when brushing your clothes. Keep woolens in plastic bags or an air-tight container. If necessary, spray baseboards and storage areas.

Centipedes and Millipedes

These little scavengers feed primarily on decaying vegetation although some species eat living plants as well.

Centipedes

Some species of these "100-leggers" can inflict a venomous bite, so be careful when handling them (Color Figure 38). Even though a bite is seldom dangerous to people, consult your physician just to be on the safe side.

House centipedes (Color Figure 16) are usually beneficial because they prey on other insects. A heavy infestation, however, can be very annoying. They are fast-moving creatures that you may see darting across the floor or walk. Originally from Mexico, they are now found throughout the United States. If your building is infested, you won't forget these strange-looking creatures in a hurry. They are about one inch long and the females have two hind legs more than twice as long as their bodies. If an animal grabs the leg of a house centipede, the leg detaches easily and the insect scurries away to safety.

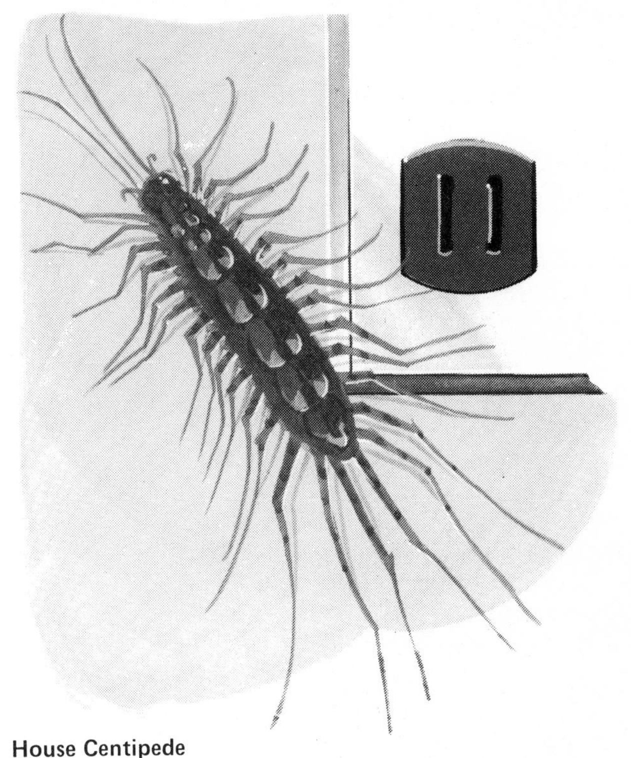

House Centipede

Their presence may indicate other insect populations that you don't know about.

You can usually control them effectively by spraying baseboards and wall and floor crevices.

Millipedes

Millipedes, also called "1000-leggers," differ from centipedes in that they have two pairs of legs per body segment while centipedes have only one pair per segment (Color Figure 39). In addition, these hard-shelled, worm-like creatures are slow-moving. Instead of darting away when disturbed, they curl up into balls.

They are often found in wooded areas or near foundations where there is decaying vegetation. New houses built on farm or forest land are particularly likely to be infested. As long as they stay outside, they present no problems, but on occasion their numbers become so large that it is almost impossible to prevent their entry. One homeowner, for example, tried to keep them out by taping the bottoms of doors. So many thousands of millipedes were migrating across the lawn, however, that they simply swarmed up the walls and entered by unguarded windows.

You can control such invasions by removing all accumulations of rocks, boards, or debris near the foundation and spraying a three-foot strip of soil around the house. If the millipede population is very large, it may be necessary to treat the whole lawn.

Saw-Toothed Beetle

Cereal Insects and Other Pantry Pests

Finding insects in your flour, noodles, or cereal is enough to make you lose your appetite fast! The number one pantry pest is the saw-toothed grain beetle (Color Figure 8). Despite its ferocious-sounding name, this insect is only one-tenth inch long. The peculiar shape of its thorax, which resembles the teeth of a saw, gives this little beetle its name. Both the larva and adult are fond of vegetable products of all kinds. They also like meats and candies.

Many other beetles and moths attack stored foods. Among the most common kitchen-invaders are flour, flat-grain, cigarette, and carpet beetles; bean, pea, and rice weevils; and Indian meal, Mediterranean flour, and Angoumois grain moths (Color Figures 9 and 10). You may spot these creatures inside packages of food or in the cracks and crevices of your pantry or cabinet shelves.

The best method of control is careful sanitation. These insects often enter the house with the groceries. Unpack your groceries and deposit the cardboard boxes outside immediately. If it is at all possible, transfer foods from boxes or paper packages to airtight containers made of glass or plastic. Look for insects as you transfer the food. It's also a good idea to vacuum your shelves at intervals.

Insecticides are usually unnecessary. If the infestation is so severe that you must spray, a light application immediately after cleaning the area is best. Remove all food and utensils to a safe distance before spraying. Never let the insecticide come into contact with food, utensils, or any surface that food touches directly.

Cigarette Beetle

Clothes Moths

Almost everyone is familiar with these destructive pests. They are best known for liking wool, but they enjoy any fabric made of animal fiber. Even your great-aunt's Victorian horsehair sofa isn't safe from them! Adult moths are yellow or gray with half-inch wingspans. The larvae are about one-half inch long and are white with dark heads. Both adults and larvae prefer darkness and a thorough brushing of the fabric combined with a one day airing in bright sunlight is usually enough to clear them out. Washing or dry-cleaning is also very effective. Store fabrics in plastic bags or other airtight containers. If necessary, spray storage areas or use mothballs.

Cluster Flies

On warm, sunny winter days, homeowners or apartment dwellers may be startled to see swarms of flies hovering outside their windows. These are not the houseflies so common during the summer, but the larger and more sluggish cluster flies. The larvae live as parasites in the bodies of earthworms. When the weather gets cool, the adults move indoors for shelter. They usually enter through windows and doors or through cracks in shingles and eaves.

Since they do not feed on human food or breed in filth as houseflies do, cluster flies are annoying rather than harmful. Control is difficult to achieve because they hide in partitions, window casements, and other inaccessible spots. Sealing off openings may keep them outside. If they do manage to get inside, impregnated resin strips are more useful than sprays. As a rule of thumb, one strip per 1,000 cubic feet (length times width times height) of room space should be sufficient. Aerosol sprays provide short-term protection, but they have little residual effect, so frequent applications will be necessary. If you have a cluster fly problem, consider spraying the outside of your house next fall.

Cockroaches

"The worst experience of my life was when I picked up my flute to practice and I saw roaches crawl out. I haven't played since."

"I had to quit my job and move out of Manhattan because I couldn't stand the roaches."

"I used to have a Buddhist-like reverence for life—catching bees trapped indoors and setting them free; refusing to destroy spider webs. But I became a maniac when I discovered my whole building was infested with roaches. I'd get up in the middle of the night, put on my shoes, and tiptoe to the kitchen. Then I'd switch on the light and stomp on as many as possible, shouting 'Kill! Kill!'"

These are just a few testimonials to the emotional impact cockroaches have on otherwise rational people. Even though man has been sharing his food and living quarters with roaches throughout history, he still hasn't learned to do it graciously.

"Friends, Romans, and Cockroaches"

Two thousand years ago when the Caesars were Johnny-come-latelies, cockroaches were firmly established in the Eternal City and throughout the Mediterranean world. Scientists believe the five major domestic species originated, like man himself, in Africa and spread out in all directions.

Dioscorides Pedanius, a Greek who served as a doctor in the Roman army during the reign of Nero (A.D. 54 to 68), believed that cockroach entrails mixed with oil and stuffed into the ear would cure earaches. Far from being a quack, Dioscorides was a highly respected medical authority whose treatise *Materia Medica* was a basic text for students for 16 centuries.

The first scientific "encyclopedia," Pliny the Elder's *Natural History,* recommended the same treatment but declared that the oil used had to be rose oil. Pliny thought of cockroaches the way we think about penicillin. Crushed cockroaches, he believed, could cure scabs, itching, swollen glands, and even tumors. This great naturalist died tragically while observing the eruption of Vesuvius that destroyed Pompeii in A.D. 79.

The Roman Empire ended, but cockroaches continued to expand their territories. When Sir Francis Drake, the famous explorer and naval officer, captured the *Philip,* a ship in the Spanish Armada (A.D. 1588), he found the decks crawling with roaches. In 1611, Danish naval records described cockroach hunts on Danish vessels. Any sailor who caught 1000 roaches received a bottle of brandy from the ship's stores. On one memorable occasion, 32,500 roaches were turned in. Captain William Bligh, immortalized by the mutiny on his ship, *The Bounty,* tried to eliminate them with boiling water. The Japanese Navy in 1905 used a variation on the Danish approach. Any sailor who captured 300 roaches received a one-day shore leave.

Life Cycle and Life Style

If you had a time machine that could carry you 265 million years into the past, you would find a very different planet. The luxuriant jungles and swamps that later produced enormous coal deposits were flourishing. Primitive land animals were evolving from the amphibians. Amid all these strange creatures you would have no trouble identifying one familiar one—the cockroach.

The cockroach fossils that have been discovered show that the remote ancestors of modern cockroaches were almost identical to their descendants. The fact that cockroaches have changed so little over this enormous span of time shows how successful they are. Today, there are more than 3,500 species around the world.

They are uncomplicated creatures that don't need much to survive. Although the cockroach prefers starchy foods, it can eat almost anything. Paper or dirty clothes are as appealing as food crumbs. If necessary, it can survive for long periods without food or water.

Cockroaches are somewhat gregarious and will cluster together as long as the food supply lasts. When hard times hit, however, each becomes an independent foraging machine, hunting by night and hiding by day. One reason cockroaches are so successful is that their survival mechanisms are so well developed. As soon as a cockroach "feels" a presence anywhere in the room (a word, a step, or even a deep breath is enough to sound the alarm), its legs start running. That's why you will find a roach when you move something. It doesn't mean to startle you—it's simply hiding. The only time you'll see one out in the open is when it's foraging and doesn't have time to make a quick getaway or isn't close enough to a hiding place.

Roaches are very prolific—they need mate only once to produce many hatches of fertile offspring. Some species, in fact, can reproduce parthenogenetically (from the Greek word *parthenos* which means virgin) without mating. All the offspring will be female.

The cockroach begins life in an egg capsule shaped like a lima bean. Each capsule contains from 16 to 40 future roaches. After hatching, it is known as a nymph. A nymph is similar to an adult but lacks wings, genitalia, and distinct markings.

The nymph becomes an adult by molting. It sheds its outer skin, or exoskeleton, 6 to 12 times, becoming a little larger each time. Oddly enough, newly molted nymphs are often eaten by other cockroaches because they are white. Until they turn dark after being exposed to air, they aren't recognized by their fellows.

Cockroaches Are Dangerous as Well as Disgusting

Cockroaches are usually unpopular because they eat anything, have filthy habits,

American Cockroach

and contaminate food and stain or damage everything they come near.

Until recently, there was no evidence that they carried diseases as well, but the latest research has proven that they transfer harmful bacteria. Among the pathogens (disease-carrying organisms) they carry are those responsible for boils, bubonic plague, diarrhea, dysentery, food poisoning, gastroenteritis, intestinal infections, leprosy, typhoid fever, and urinary tract infections. Obviously, this new information won't win them any friends!

The Five Major Domestic Species

One variety of cockroach would be bad enough, but we have to contend with five! Each species has its own habits and idiosyncrasies and it is important to identify the one that has invaded your home so you know where to spray.

American Cockroach—This large (about 1½ inches long), reddish-brown roach is also known as the palmetto bug, the water bug, and the Bombay canary (Color Figure 4). It is among the filthiest of all insects and is usually found in basements and bathrooms. Look for it near pipes, drains, and plumbing fixtures.

Brown-Banded Cockroach—This pest came to the United States from Africa by way of Cuba. First seen in Miami and Key West in 1903, it rapidly spread across the entire country. Scientists attribute this incredible dispersion to its habit of hiding and laying eggs in luggage and furniture. It is less than one-half inch long and its pale gold to brown body is marked with two brown bands across the back, giving it its nickname of the brown bandit (Color Figure 5).

It prefers heated rooms and is usually seen above ground level. Look for it in bedrooms, closets, desk and bureau drawers, behind pictures and wallpaper, and inside book bindings, clocks, and telephones. If you find one, there will be more nearby because this insect clusters with its pals.

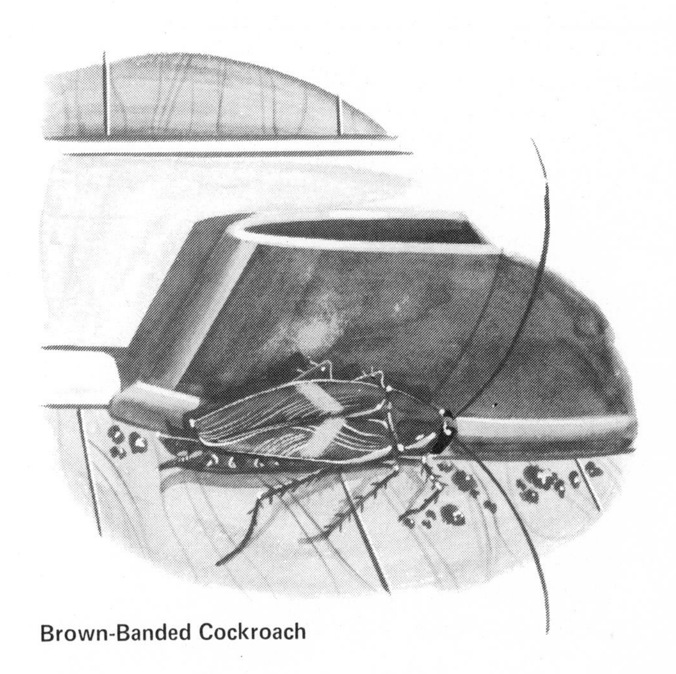

Brown-Banded Cockroach

German Cockroach—There is some argument about the origin of this pest's nickname of croton bug. Large numbers of them appeared in New York after the construction of the Croton aqueduct that provides the city with water. On the other hand, *croton* means bug in Greek.

This roach is about one-half inch long and has two dark streaks running lengthwise down its tawny body (Color Figure 2). It thrives in cooking areas and loves heat and moisture. Look for it near hot-water pipes, under sinks and the stove, and behind the refrigerator.

Oriental Cockroach—This creature's habits are similar to those of the American cockroach. You can distinguish this one by its slightly smaller size, short wings, and darker (almost black) color (Color Figure 3). It lurks in cool damp areas like basements and crawl spaces. It can live in sewers and sometimes enters the house through sewage drains. Look for it around toilets, bathtubs, and sinks. Large numbers sometimes congregate near these sources of water.

Woods Roach—This pest usually stays outdoors and seldom ventures inside (Color Figure 1). Anyone who stores wood indoors, however, may have trouble. Look for it near lumber or firewood in basement or garage.

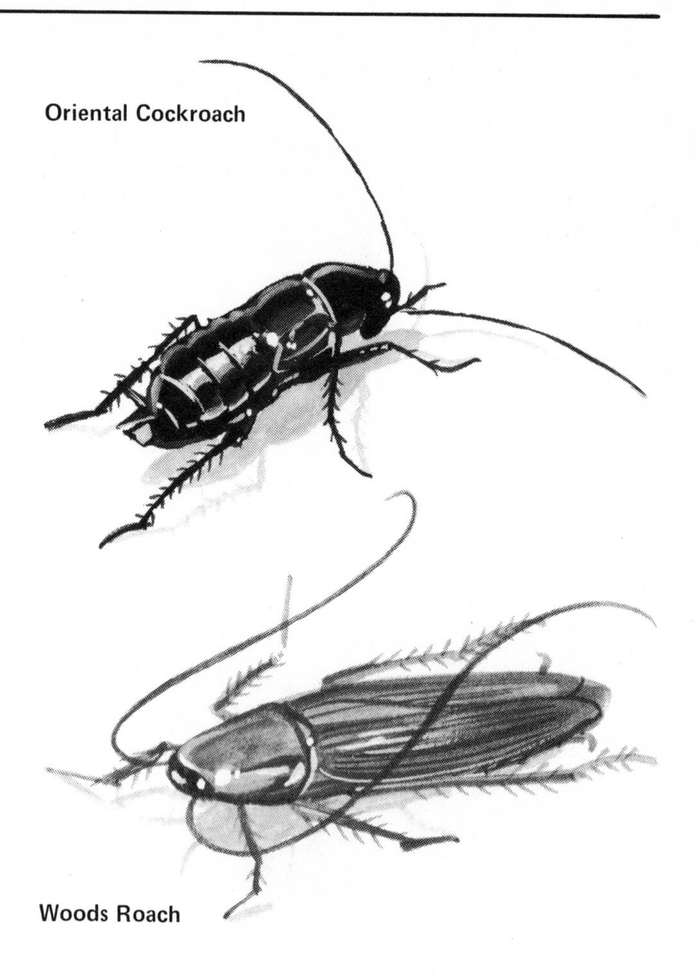
Oriental Cockroach

Woods Roach

German Cockroach

Control

Be careful when bringing supplies or second-hand furniture into the house. Look for roaches when putting food away and get rid of bags and cartons immediately. Treat furniture before bringing it inside.

If you have been invaded, locate the favorite hiding places of the roaches and treat them thoroughly. Also treat baseboards, under and behind major appliances, areas where counters and walls meet, and all pipe openings. The new pressure-sensitive, impregnated roach tapes now being marketed are not very effective and can be recommended only for very light infestations.

A Final Note of Caution

Some people become desperate enough to move when faced with cockroaches. Unfortunately, they often take their problem with them. Check books and furniture carefully before moving. Those handy cardboard boxes you got from the supermarket should also be treated before you carry your possessions to your new home. Supermarkets and grocery stores almost always have roaches.

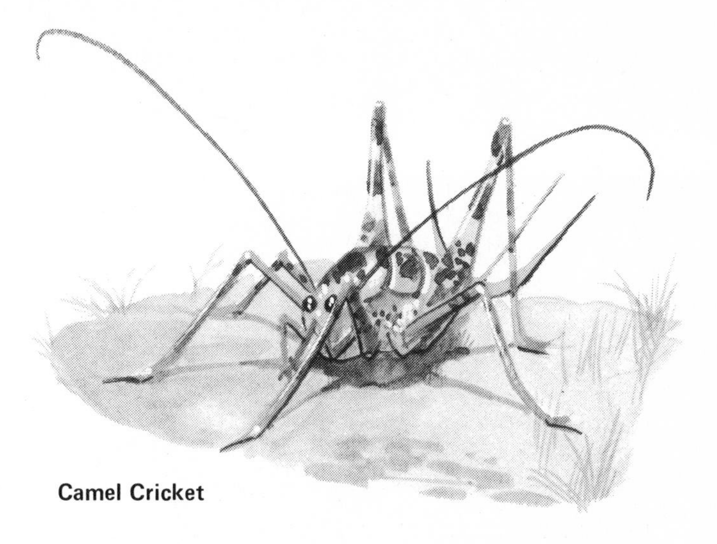

Camel Cricket

Crickets

Crickets, like so many other insects, come in a variety of forms, some more annoying than others. The cricket population varies considerably from year to year. Scientists suspect that this is because eggs are laid singly without a protective secretion. Therefore, a dry season delays or curtails hatching.

Disney movies to the contrary, crickets in the house can be a real pain. In addition to their annoying chirping, they can ruin fabrics, foods, and paper products.

Camel Crickets—Also known as cave or stone crickets, these are primarily outdoor insects (Color Figure 37). They like cool, dark, damp areas and often live under rocks and logs. Because of these habits, they may be brought indoors with the firewood. Once inside, populations grow quickly in damp cellars or seldom used fireplaces. This may be the origin of the famous "cricket on the hearth."

Field Crickets—These hardy outdoor types are not adapted to life indoors but, when their food supply dwindles and the cool evenings of late summer begin, they are apt to seek food and warmth in any convenient shelter—including your house. Field crickets are usually black and about one inch long.

House Crickets—These are the insects that cause the most damage. Yellow-brown in color and about three-quarters inch long, they prefer warm, dark places inside buildings. They hide in cracks and under objects during the day and emerge in the early evening. Adult crickets are able to fly and can jump considerable distances. Because they can climb up the sides of buildings, they have little trouble reaching upper-story apartments. Once they're inside, they set up housekeeping and begin to reproduce within the structure.

Mole Crickets—These beady-eyed creatures resemble the mole for which they are named. Their short front legs are used for digging the burrows in which they live. Unlike house crickets, they are poor jumpers and weak fliers. When the soil is flooded, they may leave their homes and make short nocturnal excursions along the surface of the ground. At these times, they may enter basements and first-floor dwellings.

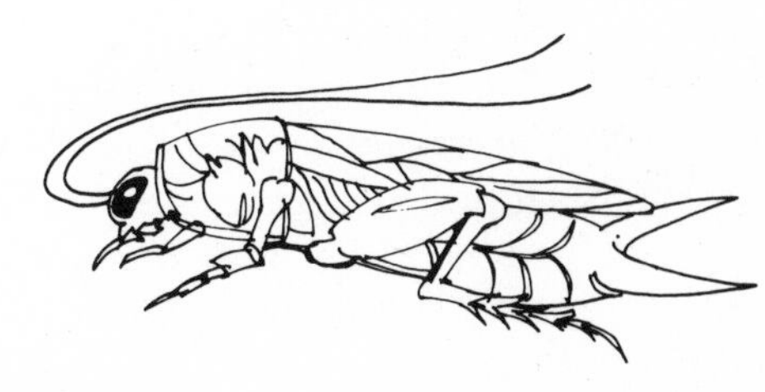

Mole Cricket

Control

Foundation spraying is the key to keeping all these crickets out of your home. If the population is very large, adjacent lawn areas may need treatment as well. If you have already been invaded, apply an insecticide to baseboards, crevices, or areas where crickets have been seen.

Drainfly

Daddy Longlegs

These harmless, spider-like creatures (Color Figure 31) feed primarily on dead insects. They are seldom a problem.

Deer and Horse Flies

These large insects are usually found around farm animals. Even city people can be bothered by them, however, because they also attack dogs and humans. Their bites can create a sensitivity that leads to lesions and high fever.

Control is difficult because the larvae are semi-aquatic. It may take almost a year before they develop into adults. During the larval stage of their lives, these flies live in mud or wet soil near lakes, streams, and marshes. A good repellent can keep them away from you.

Drainflies

These hairy, moth-like insects have an unusual triangular appearance because of the way their wings fold (Color Figure 20). They run in an irregular, hestitant manner and make short jumps when flying.

The adult deposits eggs in decaying organic matter in sewers and drains. The small, newly hatched maggots feed on the decaying material. Adult flies are usually found in damp areas such as basements, but occasionally wander into other parts of the house.

Chemical controls should be used only as a last resort. Start by cleaning out drainpipes, overflow drains, and any other standing water that might provide a breeding place. Commercial drain cleaners or boiling water poured down the drains will eliminate maggots quickly.

Dry Rot

Dry rot is caused by fungi that grow on damp wood. Damage caused by these fungi is often mistakenly blamed on termites. "Dry rot" is actually a misnomer, because a significant amount of moisture must be present in the wood in order for the fungi to cause decay. The fungi bore through cell walls by means of an enzyme they secrete. Once the cell walls are pierced, the fungi begin to feed and the wood starts to disintegrate. Look for brownish discolorations and a crumbly appearance.

Because it is difficult to control temperature and humidity, you must direct your efforts to controlling the moisture content of the wood. Many excellent wood preservatives are available. In addition, all decaying wood should be replaced. Keep lumber in a dry area above ground level. Wood scraps should be removed rather than buried.

Earwigs (see, "Lawn Pests")

Elm Leaf Beetles

Elm leaf beetles are usually considered outdoor pests, but when populations are heavy, they can drop off their favorite elm trees and crawl into your house. In most instances, a thorough vacuuming will eliminate them. If necessary, space spraying will provide good control.

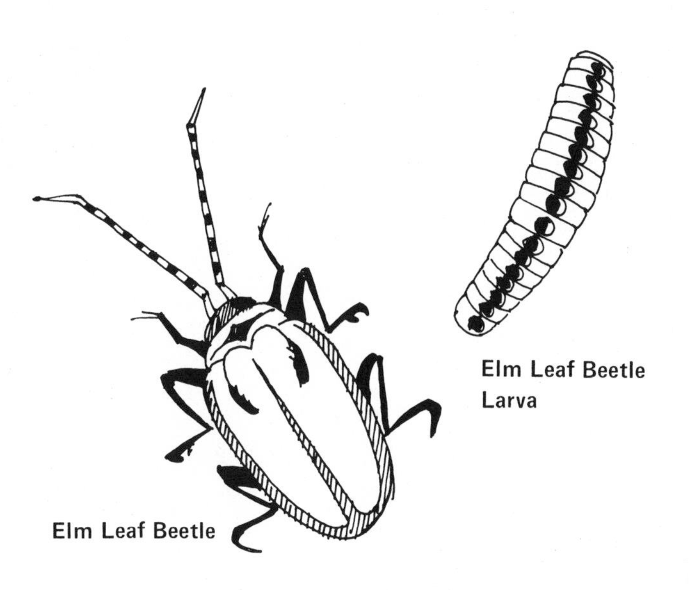

Elm Leaf Beetle

Elm Leaf Beetle Larva

Fleas

Throughout most of history, fleas were extremely dangerous because they were agents in the transmission of dread diseases like plague and typhus. Today, in modern industrialized countries at least, strict health standards, control measures, and medication have reduced these insects to nuisances.

Dogs and cats attract fleas. As long as these pets are available, they seldom attack people. If the pets are removed from the home, the flea population continues to grow and attacks any warm-blooded animal that comes along. If you take your pets on vacation with you, you may return to a houseful of hungry fleas.

Flea

Flea bites may cause intense itching, leading to secondary infections. The punctures may cause severe irritation and swelling in some people.

Adult fleas are dark brown or black insects that move swiftly (Color Figure 15). The females lay their eggs among the hairs of the animal they live on. The eggs soon fall to the ground and the larvae begin to scavenge for food. They feed on dirt and debris. Inside, the larvae live in rugs, upholstered furniture, and pet baskets; outdoors, you can find them in sandy areas beneath shrubs and flowers and in areas where pets spend their time.

Keeping your pets free of fleas is your first line of defense. Dust your pets monthly from May to October and once or twice during the colder months as well. Flea collars can be an effective substitute, providing good protection for several weeks, but some animals are allergic to them. Watch carefully for any signs of allergic reaction.

If your home becomes infested despite your precautions, spraying may be necessary. Vacuum rugs thoroughly before spraying and throw out the dust bag immediately. Treat baseboards and wall and floor crevices carefully. A light mist should be applied to rugs, beds, and upholstered furniture.

Fruit Flies

Fruit flies, also known as vinegar flies, have red eyes that suggest they've been out on an all-night binge. Because they are small enough to pass through window screens, they are often found indoors wherever fruits, uncooked food, and animal excrement are present.

Control can be difficult. A bowl of fruit left on the dining room table to tempt guests may be equally attractive to fruit flies. Good sanitation is, of course, essential. Make sure that standing water (in drain pans or basement corners, for example) is eliminated. Indoor sprays can be used to good effect.

Gnats and Midges

Gnats and midges, often called no-seeums (because of their tiny size), sandflies, or punkies, are tiny flies that can be found almost everywhere. Some are suspected of transmitting yaws and pinkeye, and all are real nuisances. They can spoil a picnic or camping trip with their vicious bites. Midge flies travel in large swarms. Populations are sometimes so thick that they create driving hazards. People in the midst of a swarm often feel suffocated and find it difficult to breathe.

As evening approaches, the midges begin to fly about looking for food, and they continue to bite all night long. A midge bite is painful enough to make you look around for a much larger creature. Because these blood-suckers are so small, they pass through window screens easily. Lights attract them, so keep your shades down or curtains drawn.

Sanitation is the first step toward controlling these pests. Make sure garbage cans have tight-fitting lids and drain standing water from gutters, pools, or anywhere else it accumulates. Spray shrubbery, tall grass, flowers, and garbage cans. Impregnated resin strips fastened to the lids of the cans are also helpful. Use space sprays indoors if necessary.

Houseflies

Every time you sit down to eat during the summer, at least one fly seems to appear. You aren't depriving anyone else, however, for there are more than enough flies to go around. Flies reproduce so rapidly that scientists say that two flies mating in April would have 191 quintillion (191 followed by 18 zeroes!) descendants by the end of August. If they all survived, which fortunately for us they don't, the entire earth would have a 47-foot covering of flies.

In addition to being prolific, houseflies are also filthy. Their eggs are deposited in decaying flesh or garbage or in excrement. The adult flies ingest decaying liquids and then, attracted by warmth or the smell of cooking food, head indoors where they land on any exposed food. Besides being disgusting, these insects can be a health hazard. An adult fly (Color Figure 18) may host 250,000 bacteria. These bacteria can cause all sorts of unpleasant diseases, including cholera, typhus, parasitic ailments such as tapeworms, and gastrointestinal illnesses.

Housefly

Control can be difficult, especially if the flies breed elsewhere and then migrate to your property, but it is essential. Flies are a serious public health problem. You can discourage them by fastening impregnated resin strips to the lids of garbage cans and keeping lids tightly fastened. All animal droppings should be cleaned up immediately. If necessary, supplement sanitation measures with insecticides.

Larder Beetles

One of the more troublesome domestic pests is the larder beetle which, as you might guess from its name, lives on stored foods, including meats and cheeses. If you see this small (about one-quarter inch long), black creature scurrying around, look for a dead animal nearby. The larder beetle is often attracted by a dead mouse in the wall, a dead squirrel in the chimney, or some other little creature that became trapped and died. After removing the body (if you find one), treat baseboards, crevices, and the area where the body was found.

Lice

Lice are unpleasant to talk about and even worse to experience, but they are a growing problem and must be discussed.

Lice infestations are generally more common during the winter or when people are crowded together for long periods. However, they may make an appearance at any time. The first signs of infestation are intense itching, inflammation, or a crawling sensation of the skin. Lice bites, though scarcely felt at the time, can become irritated. Scratching can lead to secondary infections. There are three major species of lice that attack humans (Color Figure 11).

Head Lice—The head louse, as its name implies, lives among the hairs on the head, moving to the skin only to feed. They are usually transmitted through the use of infested combs, brushes, wigs, hats, or towels.

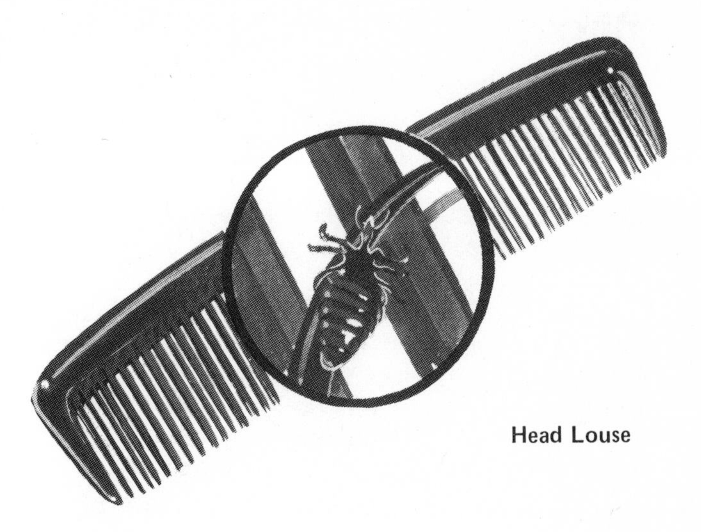

Head Louse

Body Lice—Very similar in appearance to the head louse, the body louse lives in body hair. If you look closely, you may see louse eggs (nits) attached to hairs close to the skin. If possible, check with a magnifying glass. A close look may reveal minute, grayish specks. The body louse can survive in moist clothing or bedding for a month or more without feeding.

Crab Lice—Crab lice are usually found in the hairs of the armpits and legs and in the pubic region. They can be distinguished from head and body lice by their shorter, heavier bodies. Both crab and body lice are transmitted by clothing, bedding, and intimate contact.

Control

If you have a problem with lice, your family physician should be consulted at once. A light insecticide dust may be applied (do not use dust near your eyes). Thorough, daily cleansing of the body and clothing is essential.

Millipedes (see, Centipedes and Millipedes)

Mites

Scientists guess that between 50% and 70% of the earth's people are infested with mites—but most of them don't know it. "Infested" is perhaps too strong a word because it usually implies a serious problem and most individuals are not bothered by these tiny parasites.

In severe cases, a person may feel an itching or crawling sensation of the skin followed by a pinprick. Other symptoms may include hive-like welts, scab formation, or skin sloughing. Because these same symptoms can be caused by a variety of skin disorders totally unrelated to mites, it is essential to identify the mite for a positive diagnosis. Mites come in various colors—black, gray, and red mites are common—but most are oval in shape. The follicle mite, however, is shaped like a worm. They are piercing and sucking creatures related to spiders and ticks. Most of the mites that attack people are microscopic or barely visible. There are a number of species often found on man.

Bird Mites—Wild birds often build nests in protected areas near houses. A window ledge next to an air conditioner is a favorite spot. Because they are so tiny, bird mites can pass through screens easily. Once inside, they may take up residence on you! Bird mites, fortunately, are unable to survive very long on human hosts.

Cheese Mites—These creatures, also known by the very descriptive name of grocer's itch mites, feed on flour, sugar, meats, dried fruits and, of course, cheese. If you work with any of these products or handle them extensively, you are likely to be affected.

Chiggers—Chiggers are outdoor creatures most likely to bother campers. They live in grasses, weeds, and various sorts of vegetation. If you come in contact with this vegetation, the mites may transfer themselves to you. They then tunnel into a skin pore or hair follicle, usually around the waist or groin area where clothing is snug.

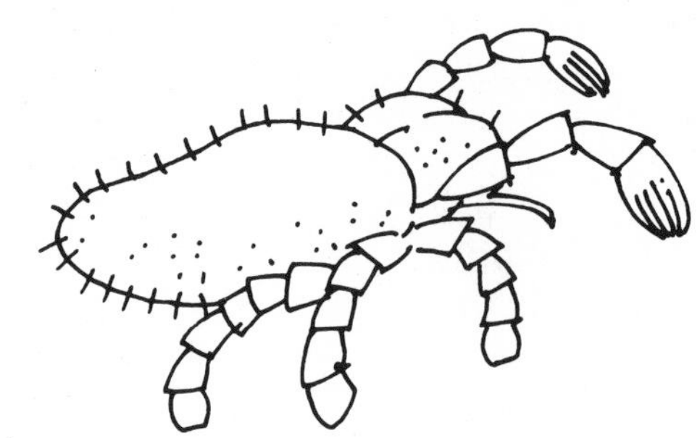

Chigger

Clover Mites—These pesty creatures do not bother people but they can overrun homes. A single room may contain up to 220,000 clover mites, all busily crawling around. When crushed, these tiny red-black specks (about the size of a pencil dot) leave red marks that stain walls and fabrics. Clover mites usually spend the winter under the bark of trees, but they have also been found in the cracks of building exteriors. When the warm spring weather arrives, the eggs hatch and they become a messy problem.

Follicle Mites—These worm-shaped mites live comfortably on humans, usually in hair follicles, the sebaceous glands of the face, or in ear wax. They are extremely common, but fortunately only a few people are sensitive to them.

Harvest Mites—Also known as straw itch mites, harvest mites are likely to be encountered by people who work with straw, hay, cotton and other crops. They are particularly abundant during the harvest season. Because they also feed on insects, they are sometimes considered beneficial.

House Dust Mites—These mites are the ones most likely to infest houses. They are usually found in pillows, mattresses, and upholstered furniture stuffed with natural materials. On humans, they concentrate on the head, tunneling along the hairline, near the eyebrows, and around the nose.

Human Itch Mites—These mites tunnel into

the skin of the hand and wrist. One of their favorite spots is the skin between the fingers. Also called scabies mites, these creatures are similar to the mites that cause mange in domestic animals and pets.

Tropical Rat and House Mouse Mites—As their names suggest, these mites are found primarily on rodents. Because they can transmit rickettsial pox, a typhus-like disease, they are considered dangerous. Immediate control measures are necessary.

Control

Whenever mites are suspected, a dermatologist should be consulted. If you can feel movement on your skin, moisten a piece of sterile cotton with rubbing alcohol and press it to the skin for a minute or two. The dermatologist can then examine the cotton in order to reach an identification. A germicidal soap used as a shampoo provides some temporary symptomatic relief, but does not kill embedded mites or eggs. If you are going to spend time in an area suspected of being infested, use a good repellent on your clothing and exposed parts of your body. A warm, soapy bath or shower as soon as possible after exposure is a good idea.

If you see mites in the house, a vacuum cleaner is an effective means of eliminating them. (A small amount of the dust picked up by the vacuum can be brought to the dermatologist or entomologist for mite identification.)

To keep them from entering, spray the outside walls of the house up to window level and all grassy areas within ten feet of the building. These sprays, while effective, may leave a white residue on dark-colored houses, so test spray an inconspicuous portion of the wall first. Removing grass and weeds along the foundation will also help. You can replant this 18-inch-wide strip with flowers that do not attract clover mites. Zinnias, marigolds, chrysanthemums, and roses are all excellent for this purpose.

Mosquitoes

Most people, unaware of the grave menace to public health posed by mosquitoes (Color Figure 14), regard them as simple nuisances. However, the Center for Disease Control in Atlanta, which serves as a national clearing house for disease-related information, warns that encephalitis, a mosquito-borne disease with a high mortality rate, is becoming more widespread. Mosquitoes also transmit a number of other illnesses, some of which regularly devastate underdeveloped nations. In Bangladesh, for example, malaria kills so many people that deaths due to malaria are often attributed to "natural causes" rather than disease. Many of the diseases are not an immediate problem in the United States, but the threat of future epidemics cannot be dismissed lightly.

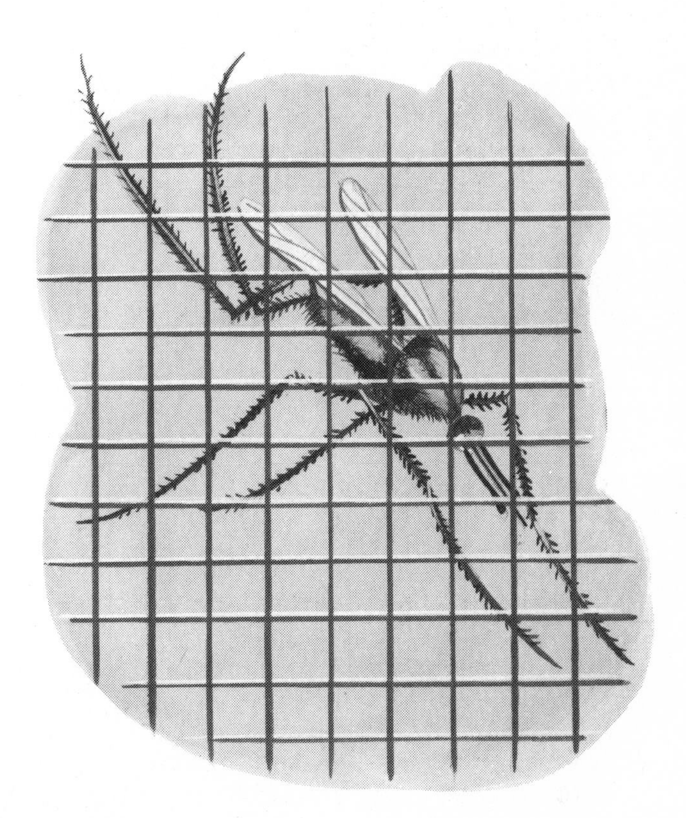

Mosquito

Community Programs

Local governments are aware of these potential dangers and community mosquito control programs are common. The cumbersome thermal fogging units that used to lumber down the streets emitting dense fogs have given way to sophisticated ultra-low-volume (ULV) units. ULV units break concentrated insecticides down into extremely fine particles that adhere readily to foliage. These units provide improved control while reducing air pollution. They also use far less insecticide than the old-fashioned foggers.

Although spraying can reduce the number of adult mosquitoes and thus provide temporary relief, effective control depends on the elimination of breeding sites. Areas containing stagnant water should be drained whenever possible and other breeding sites treated in late winter or very early spring before hatching begins.

What You Can Do

If your community has a control program, your task will be much simpler. If it does not, you can work with your neighbors to provide good control in your vicinity.

Drain standing water in eave troughs, old tires, children's pools, and anywhere else you can think of. If you have a birdbath in the backyard, change the water every three days. Fill in low-lying areas in the yard to improve drainage. These measures will keep the larva population down.

To control adult mosquitoes, spray tall grass and shrubs regularly and eliminate weed patches. Impregnated resin strips will get rid of insects that manage to get inside. Do not hang these strips in kitchens or food-handling areas and make sure that all strips are high enough to be out of the reach of children.

A good repellent applied to exposed areas of the skin will keep them away while you're outside. Those people with severe allergic reactions to mosquito bites should carry medication (available on prescription) all summer.

Because mosquitoes are so annoying, many control gimmicks are now on the market. Despite their claims, they are ineffective. Among the most popular are hand units that emit a high-pitched sound and "black lights."

In a recent test, the sound unit was turned on and placed in a cage full of mosquitoes. Far from begin repelled, many insects landed directly on the unit. Black lights proved equally useless. They are supposed to be placed 20 feet away from the spot where you are sitting. In theory, the lights attract the mosquitoes and keep them from bothering you. A test showed, however, that the number of mosquitoes attacking people was not affected by the light.

Picnic Beetles

As their name implies, these beetles are delighted by outdoor meals and think nothing of crashing your parties. People have been enjoying picnics for thousands of years, and, no doubt, the ancestors of your picnic beetles did too. These shiny black creatures, easily identified by the four prominent orange or yellow spots on

Picnic Beetles

their backs, are attracted by the aroma of food and make their appearance almost immediately (Color Figure 67). If there are enough of them around, you may find it difficult to enjoy your meal.

These irksome little pests spend the winter in decaying fruit or vegetables or other rotting vegetation. Look for them on plants after they emerge in the spring. Under normal conditions, they do not attack healthy plants but seek out wounded or damaged areas on growing plants.

Controlling the adult beetles is simple. Harvest your garden crops before they become overripe and dispose of all spoiled fruits and vegetables quickly. Spray grassy areas and garbage cans to discourage them. A second treatment four or five days later may be necessary.

Pillbugs (see, "Lawn Pests")

Powder-Post Beetles

The term powder-post beetle includes a number of wood-boring species. Different species live in different sorts of wood—some preferring softwoods and others hardwoods, some seasoned and others unseasoned woods. Although they rarely cause enough damage to weaken large timbers structurally, they can be serious pests when they infest joists, flooring, lumber, furniture, or other wood products.

The adult beetles are small, hard-shelled, and brownish in color. They damage wood by tunneling, leaving fine powdery sawdust piles in their wake (Color Figure 27). These little heaps of sawdust are often the first indication of infestation.

Powder-post beetles, unlike termites, do not need soil to survive. The females lay their eggs in the wood itself. Some species require bark, but others use wood pores or old holes in the wood.

Because they can re-infest the same wood, thorough saturation is needed to

Powder-Post Beetle Damage

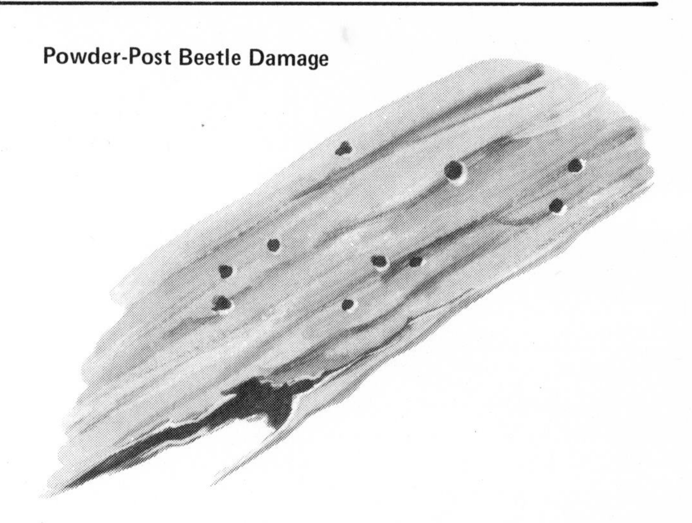

achieve control. A wood preservative with a strong repellent odor can discourage them from returning.

Scorpions

Although there are more than 40 species of scorpions within the United States, only two have potentially fatal stings. These two neurotoxin-producing scorpions have been found only in Arizona. Other species capable of inflicting painful but nonlethal bites are found throughout the South. Perhaps the most common of these is the striped scorpion, which receives its name from the broad dark bands on its back.

Scorpions are not insects but arachnids (the same group to which spiders belong). These creatures may grow to be five inches long and are in most instances nocturnal (Color Figure 33). They feed on insects and spiders which they trap with their large claws and then kill by stinging. Offspring are born alive rather than hatched from eggs and may take years to reach maturity.

As a rule, scorpions do not attack people, but they are quick to sting when disturbed. They usually lurk under boards or in rock and trash piles. Scorpion stings are as painful as those of bees and wasps and normally result in local swelling and skin discoloration.

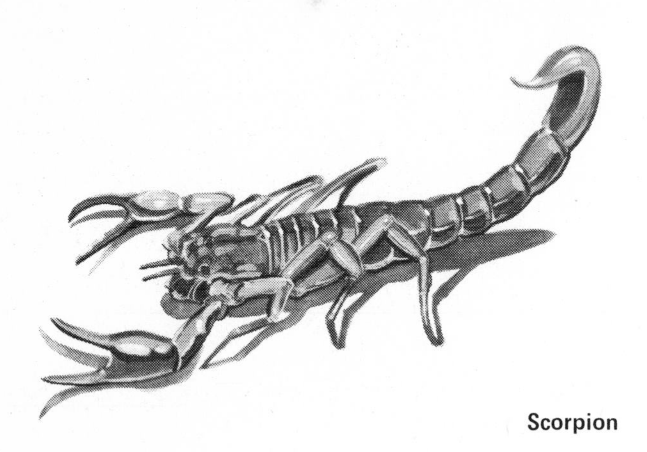

Scorpion

The northern states produce pseudoscorpions, also known as book scorpions. These are very small creatures (about the size of an ant) with flat, oval bodies and large claws. They are generally found under bark or leaves, between the pages of books, or clinging to larger insects.

Controlling scorpions is not difficult, but it does require care. Remove attractive accumulations of lumber, brick, wood, or other materials in which scorpions can live. If you suspect their presence, it's a good idea to wear heavy gloves to protect your hands from stings. Spray the foundation, baseboards, crevices, and any other spots where they may be hiding.

Pseudoscorpions usually do not require special control measures because they are seldom numerous. If necessary, you can use the methods suggested for controlling scorpions.

Silverfish

These small (one-half inch long), torpedo-shaped insects seem to be more noticeable during the winter. They are usually detected in a bathtub or sink, so most people think they enter the house through the drains. In fact, they live in more protected places and simply become trapped in sinks and tubs because they can't crawl back up the slippery porcelain.

Silverfish pass through three stages—egg, nymph, and adult—during their two- to four-week life cycle. There are normally several generations each year. Adults have long antennas and three long tails, each almost as long as their bodies. Thin scales, which are easily rubbed off, cover their bodies (Color Figure 19). The immature silverfish closely resemble the adults but are smaller.

They are considered nuisances because they damage book bindings, papers, fabrics, and wallpaper. They can cause serious problems in offices and libraries because of their taste for paper.

Silverfish can be difficult to control because they hide in secluded areas—inside stacks of paper or clothing, under insulation, or between partitions. Keeping stored papers and fabrics to a minimum will help. Insecticide sprays can be applied to baseboards, pipe openings, and any areas where silverfish have been seen. A second treatment two weeks later may be needed.

Silverfish

People Pests and Plant Pests

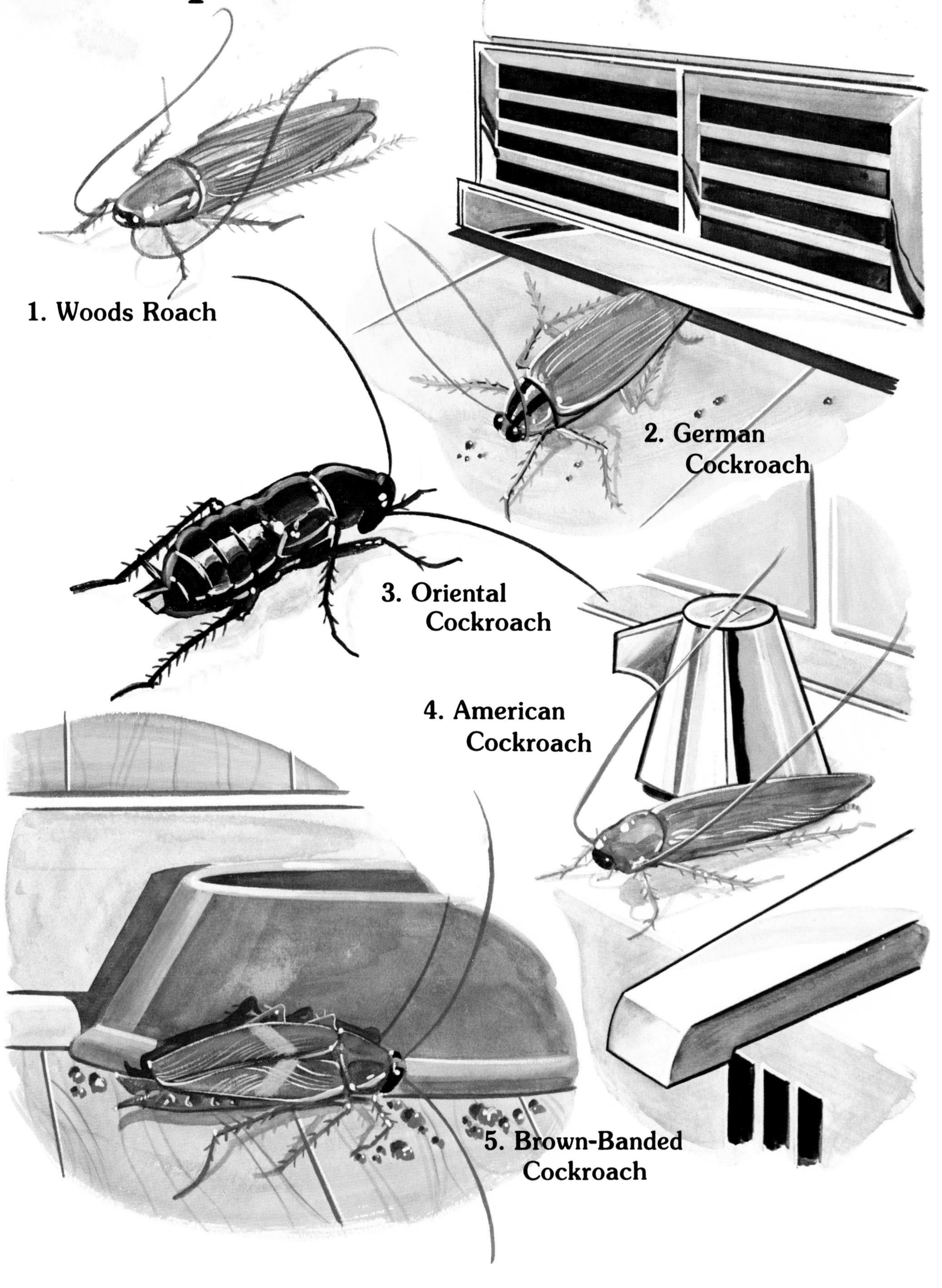

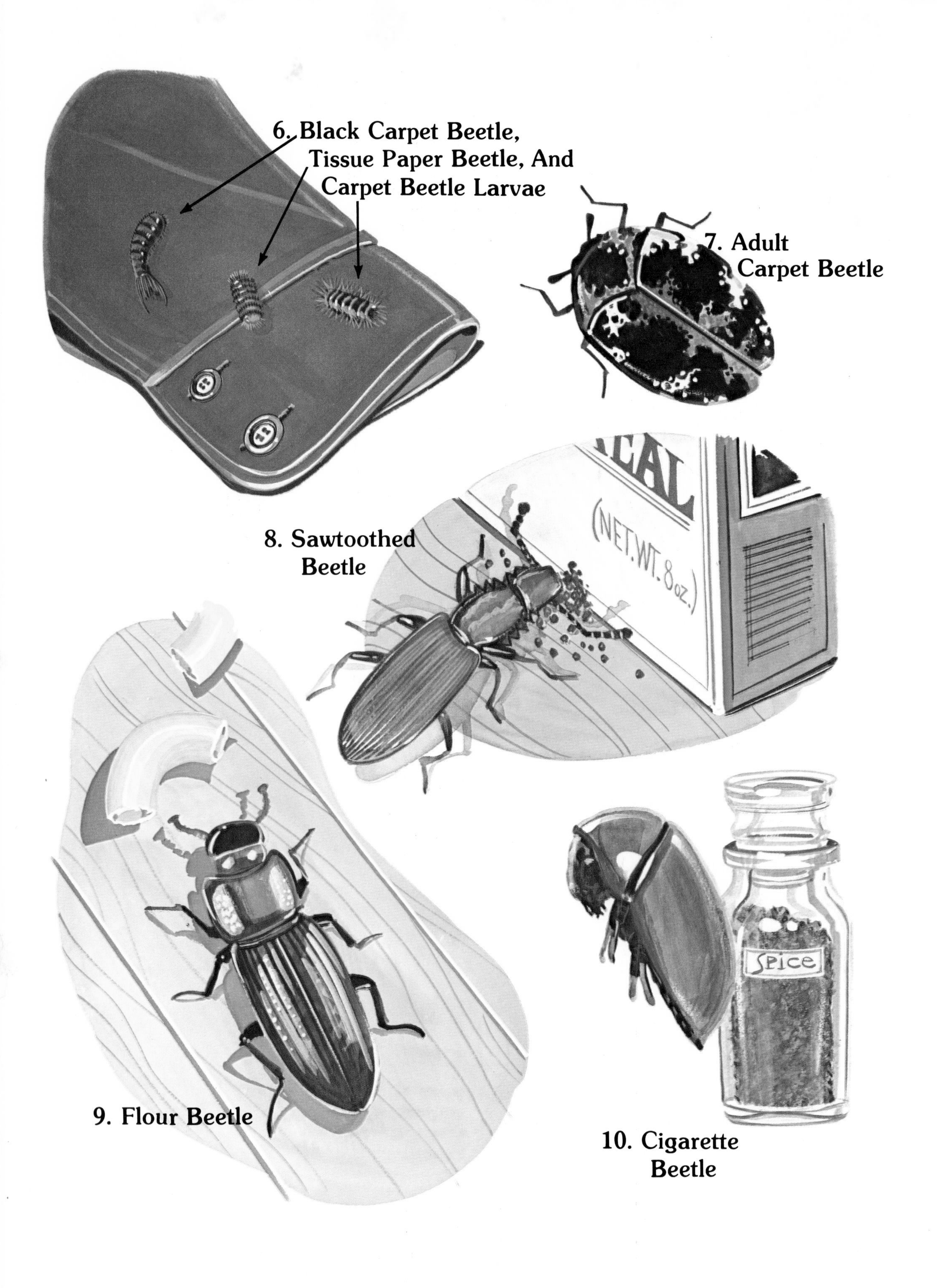
6. Black Carpet Beetle,
Tissue Paper Beetle, And
Carpet Beetle Larvae
7. Adult
Carpet Beetle
(NET. WT. 8 oz.)
8. Sawtoothed
Beetle
9. Flour Beetle
SPICE
10. Cigarette
Beetle

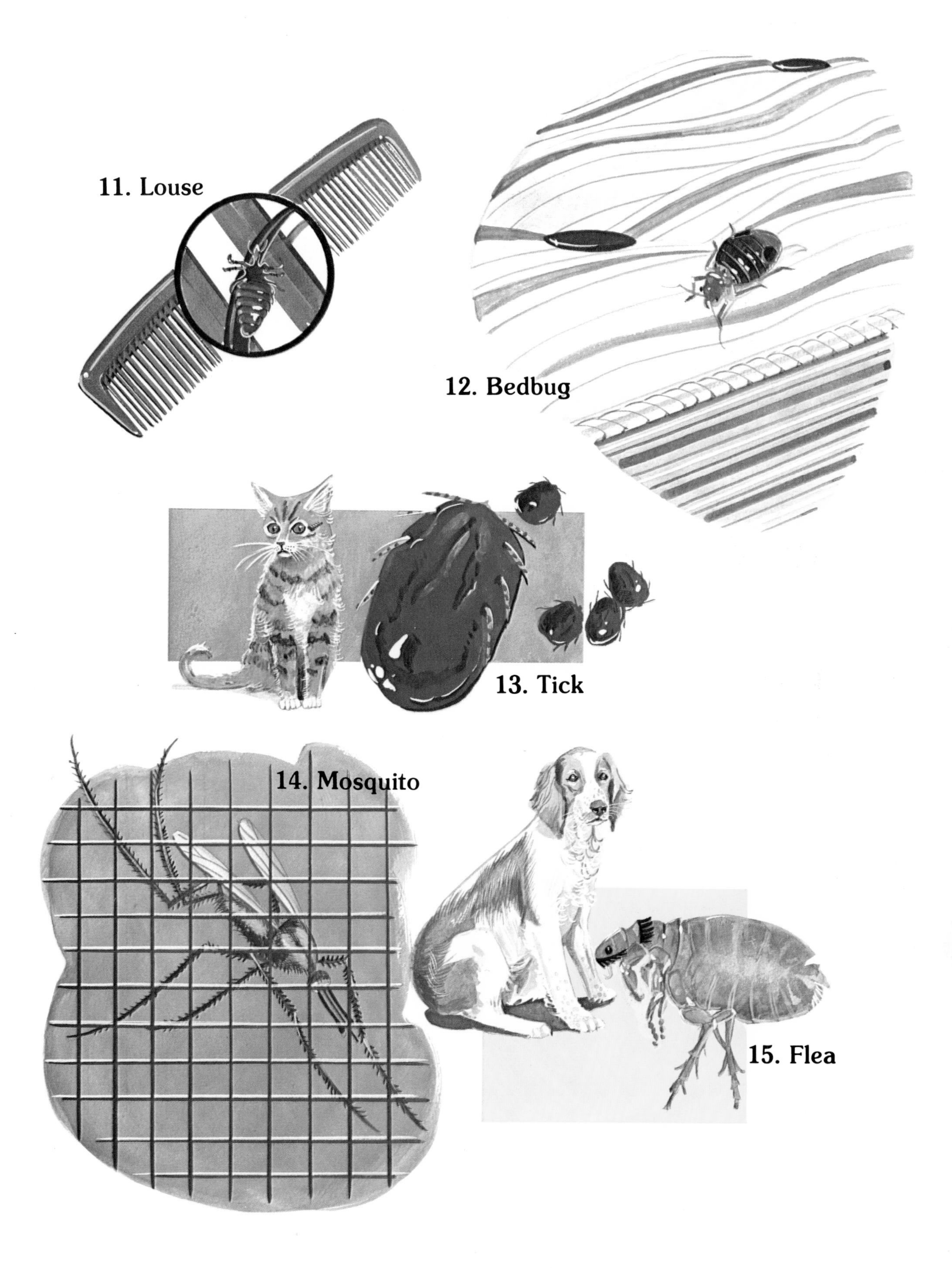
11. Louse
12. Bedbug
13. Tick
14. Mosquito
15. Flea

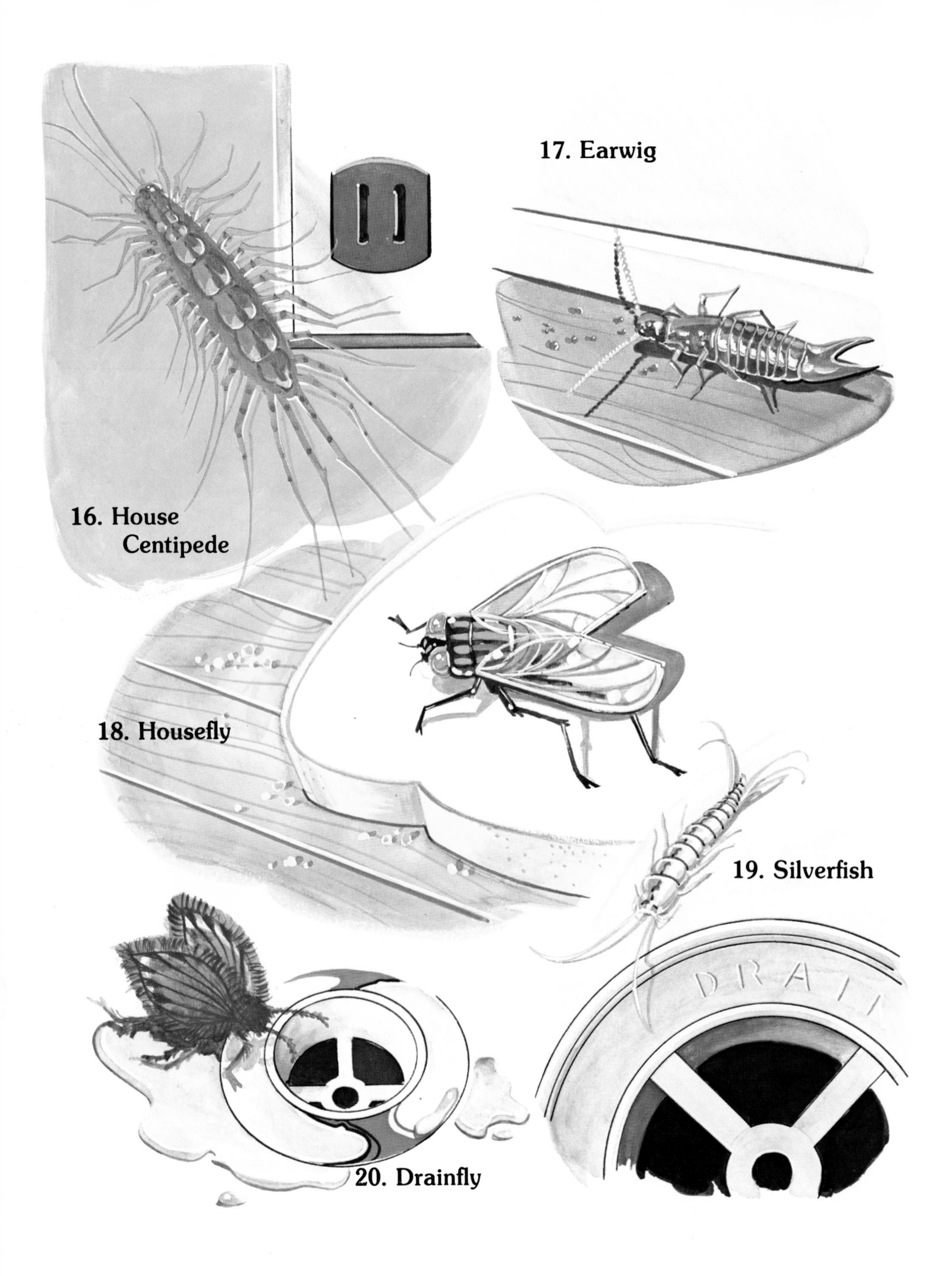
16. House Centipede
17. Earwig
18. Housefly
19. Silverfish
20. Drainfly

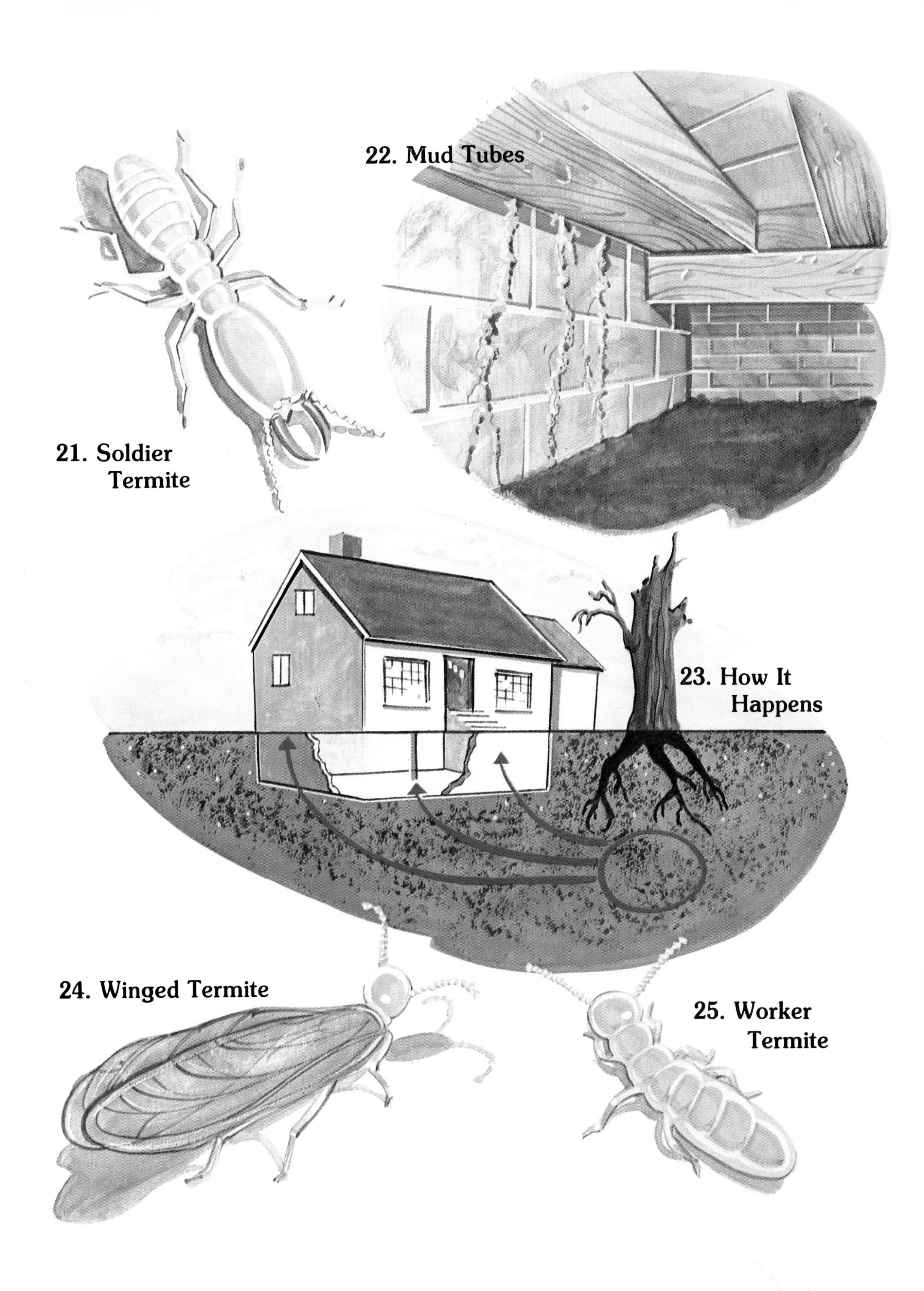
22. Mud Tubes
21. Soldier
Termite
23. How It
Happens
24. Winged Termite
25. Worker
Termite

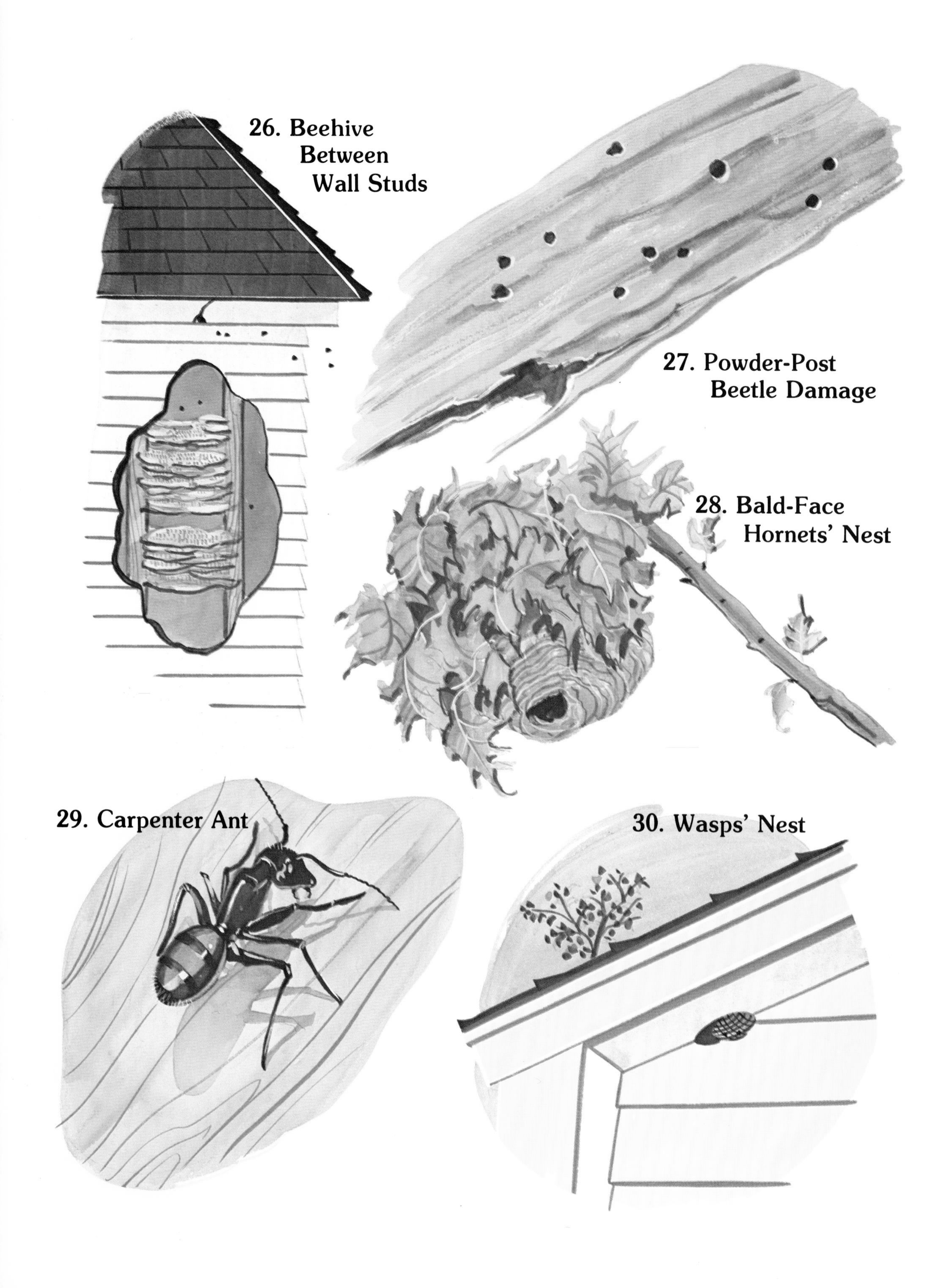
26. Beehive
Between
Wall Studs
27. Powder-Post
Beetle Damage
28. Bald-Face
Hornets' Nest
29. Carpenter Ant
30. Wasps' Nest

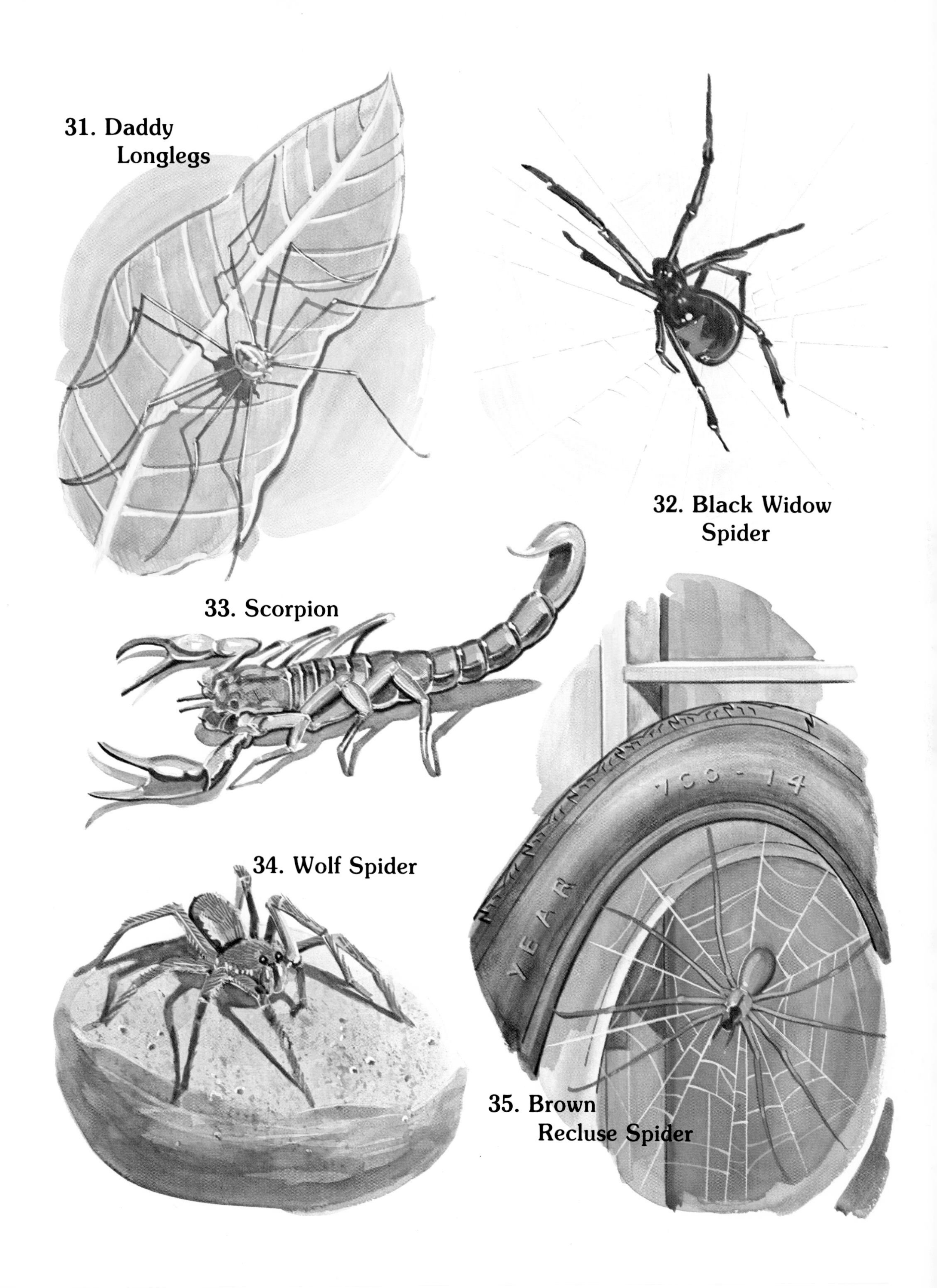
31. Daddy Longlegs
32. Black Widow Spider
33. Scorpion
34. Wolf Spider
35. Brown Recluse Spider

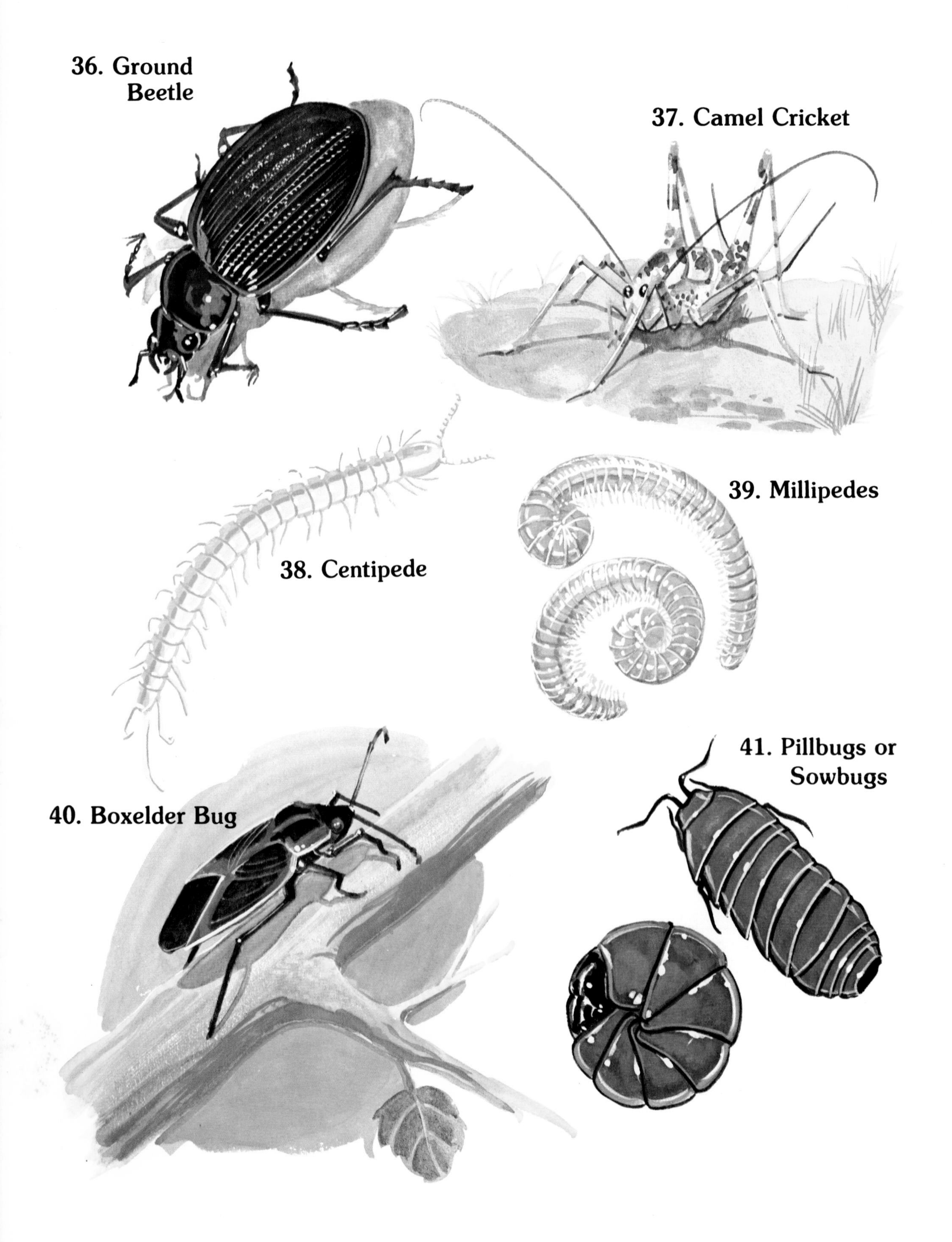
36. Ground Beetle
37. Camel Cricket
38. Centipede
39. Millipedes
40. Boxelder Bug
41. Pillbugs or Sowbugs

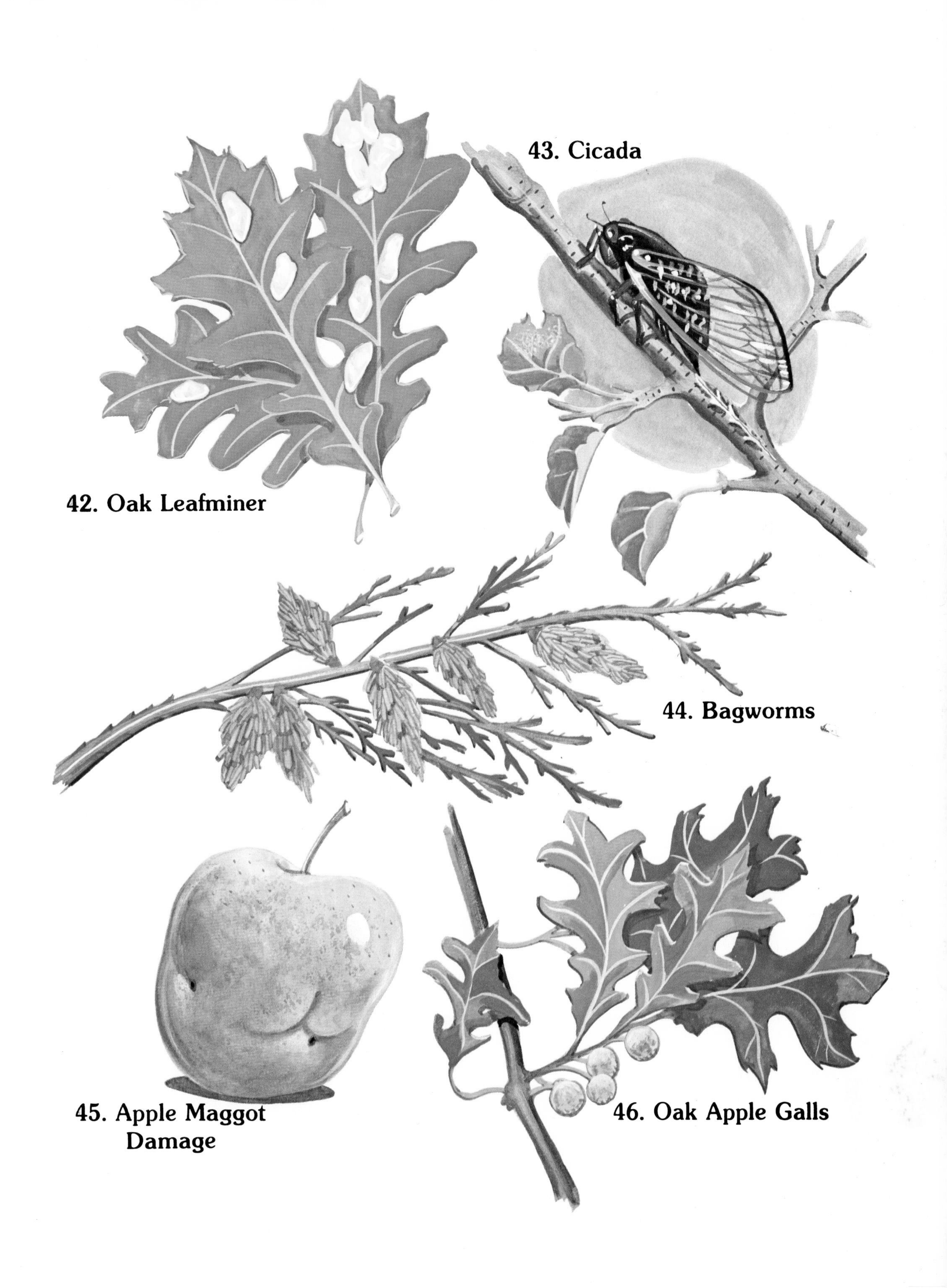
42. Oak Leafminer
43. Cicada
44. Bagworms
45. Apple Maggot Damage
46. Oak Apple Galls

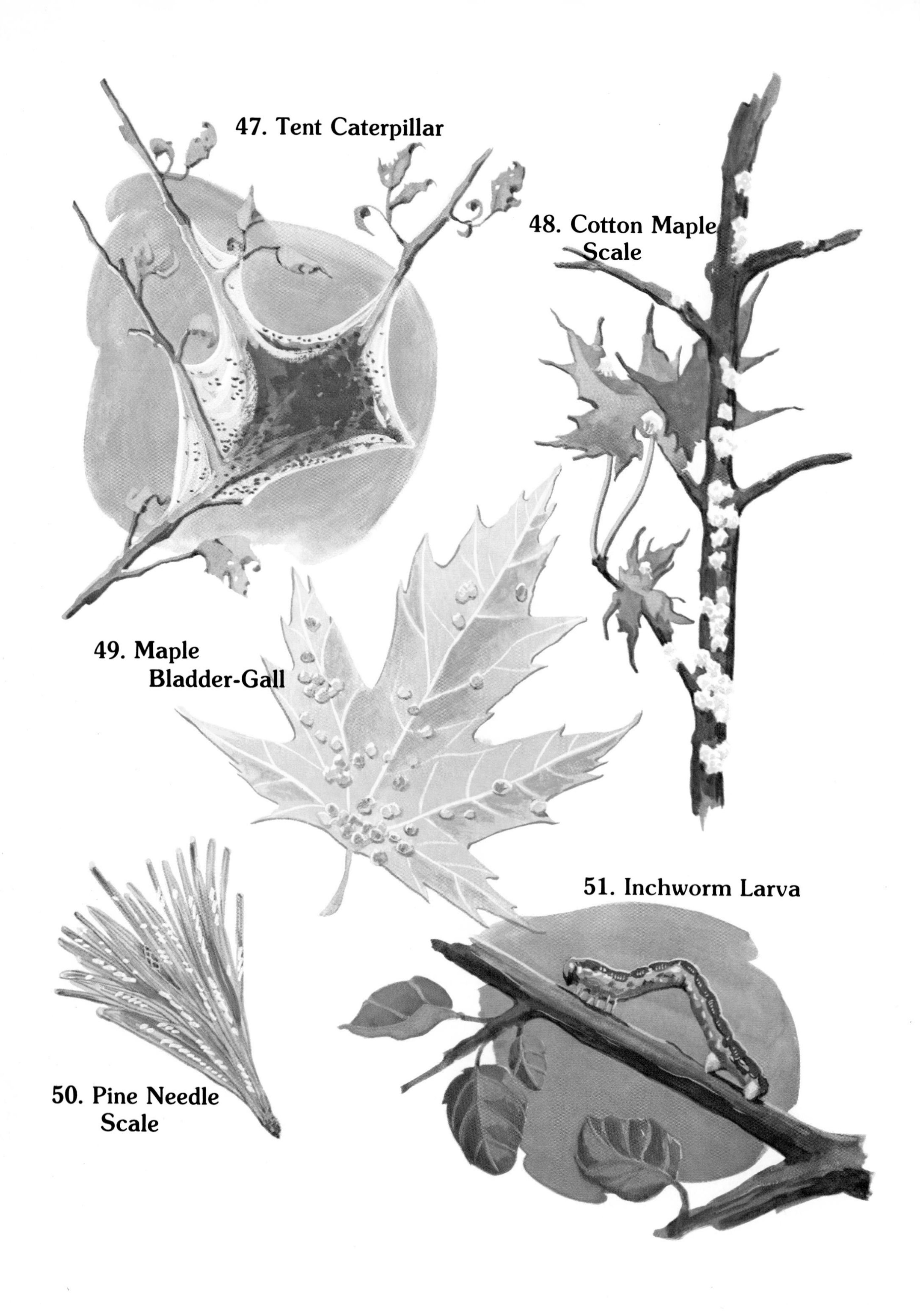
47. Tent Caterpillar
48. Cotton Maple Scale
49. Maple Bladder-Gall
50. Pine Needle Scale
51. Inchworm Larva

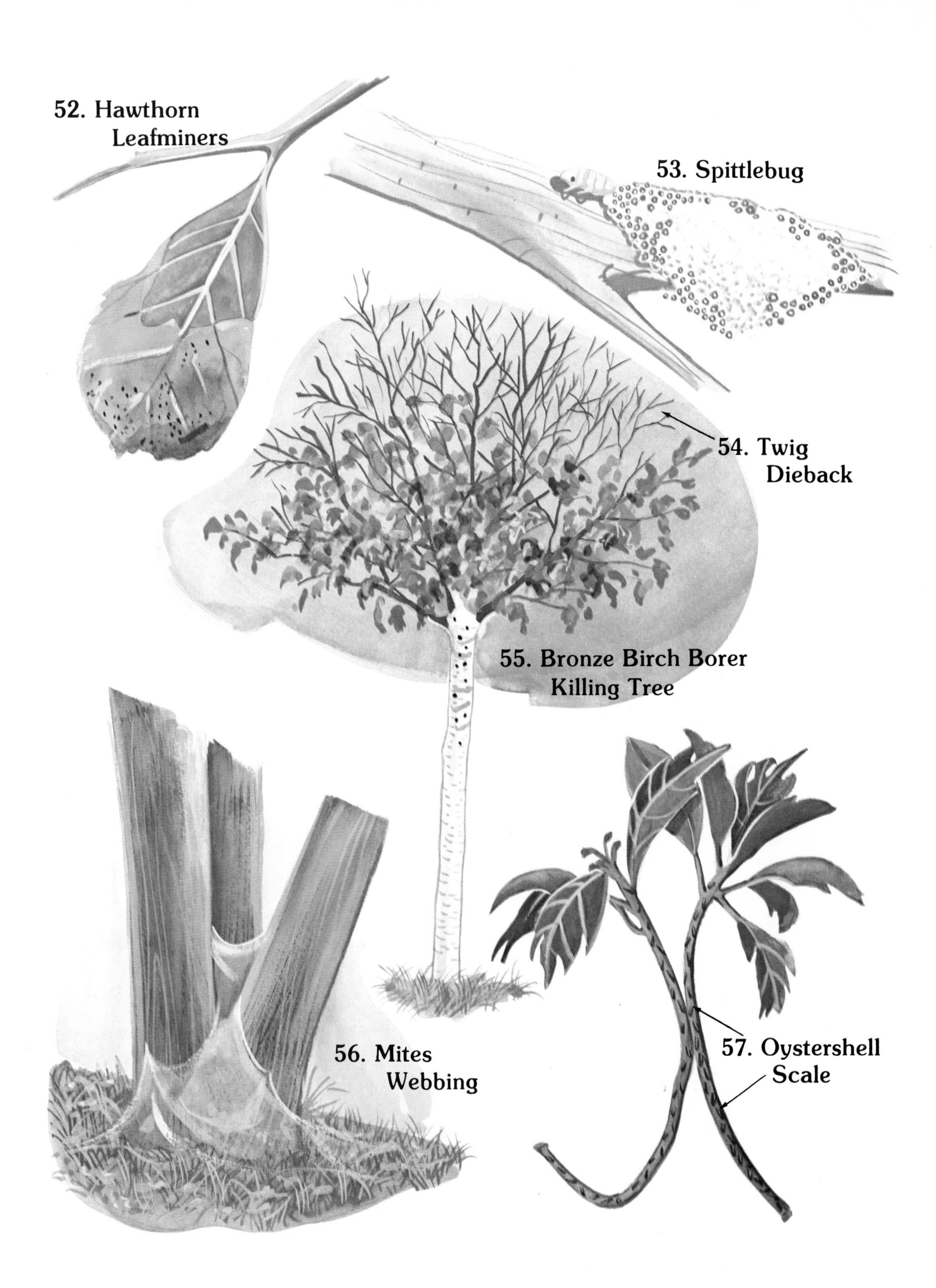
52. Hawthorn
Leafminers
53. Spittlebug
54. Twig
Dieback
55. Bronze Birch Borer
Killing Tree
56. Mites
Webbing
57. Oystershell
Scale

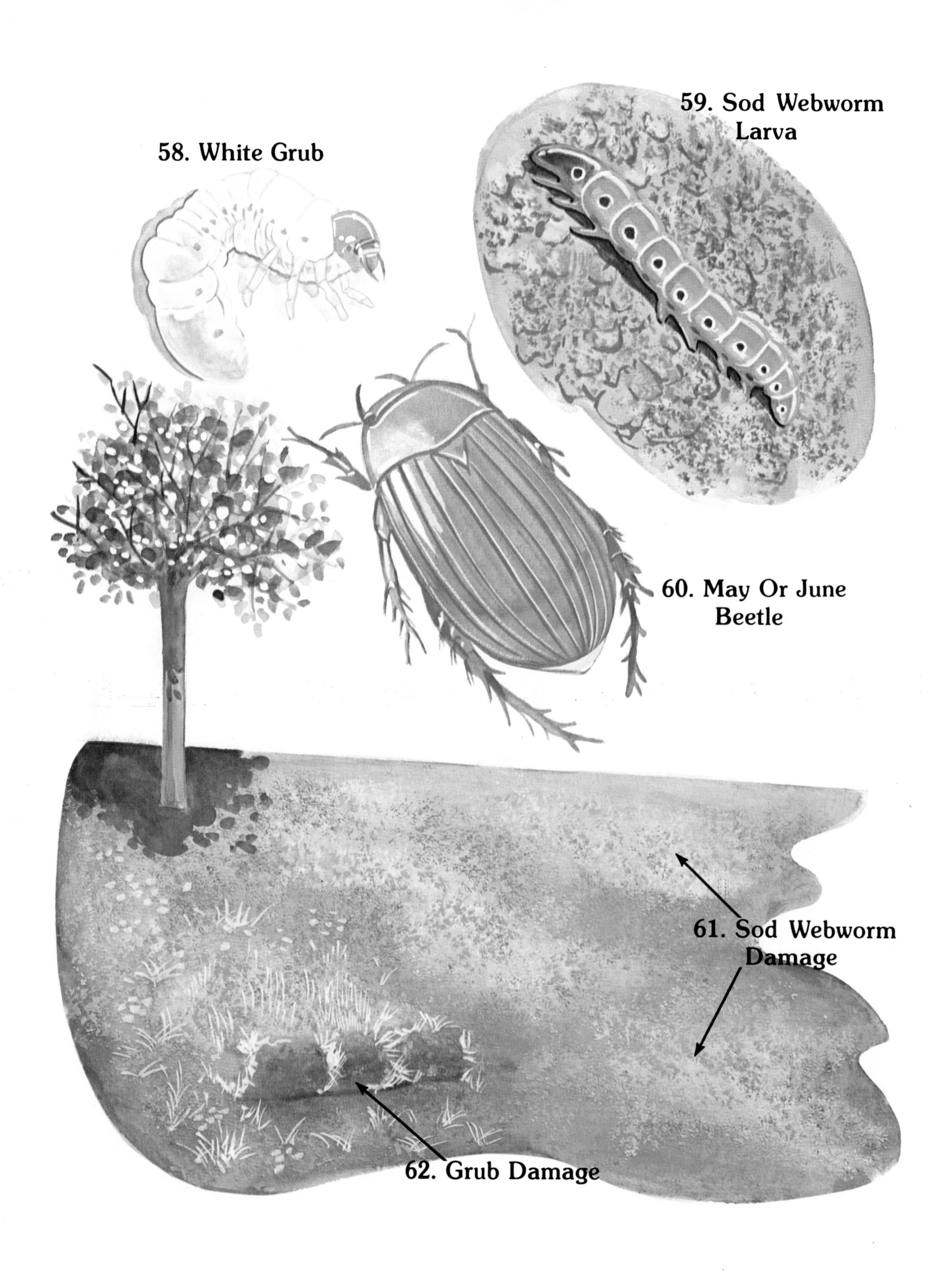
58. White Grub
59. Sod Webworm Larva
60. May Or June Beetle
61. Sod Webworm Damage
62. Grub Damage

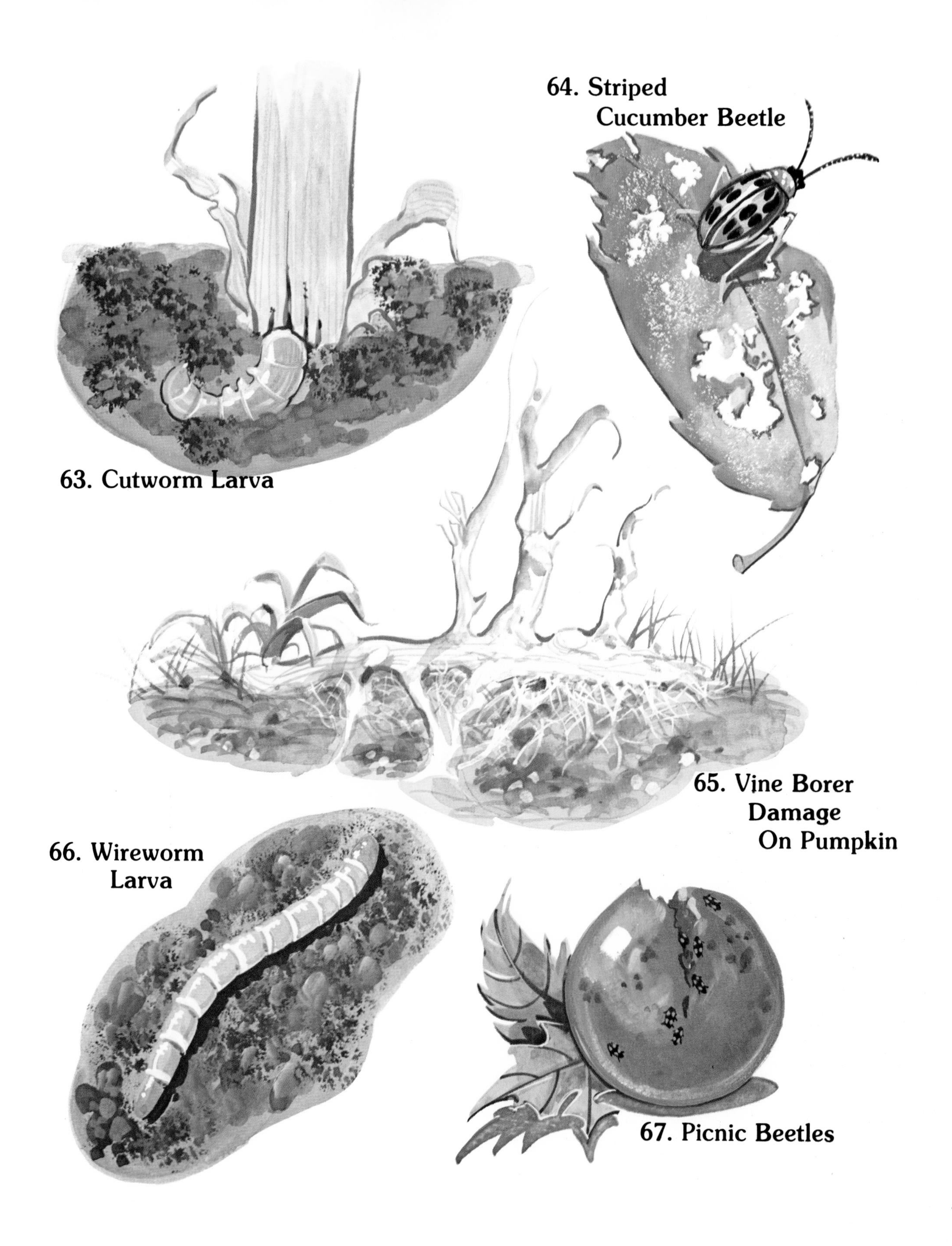
63. Cutworm Larva
64. Striped Cucumber Beetle
65. Vine Borer Damage On Pumpkin
66. Wireworm Larva
67. Picnic Beetles

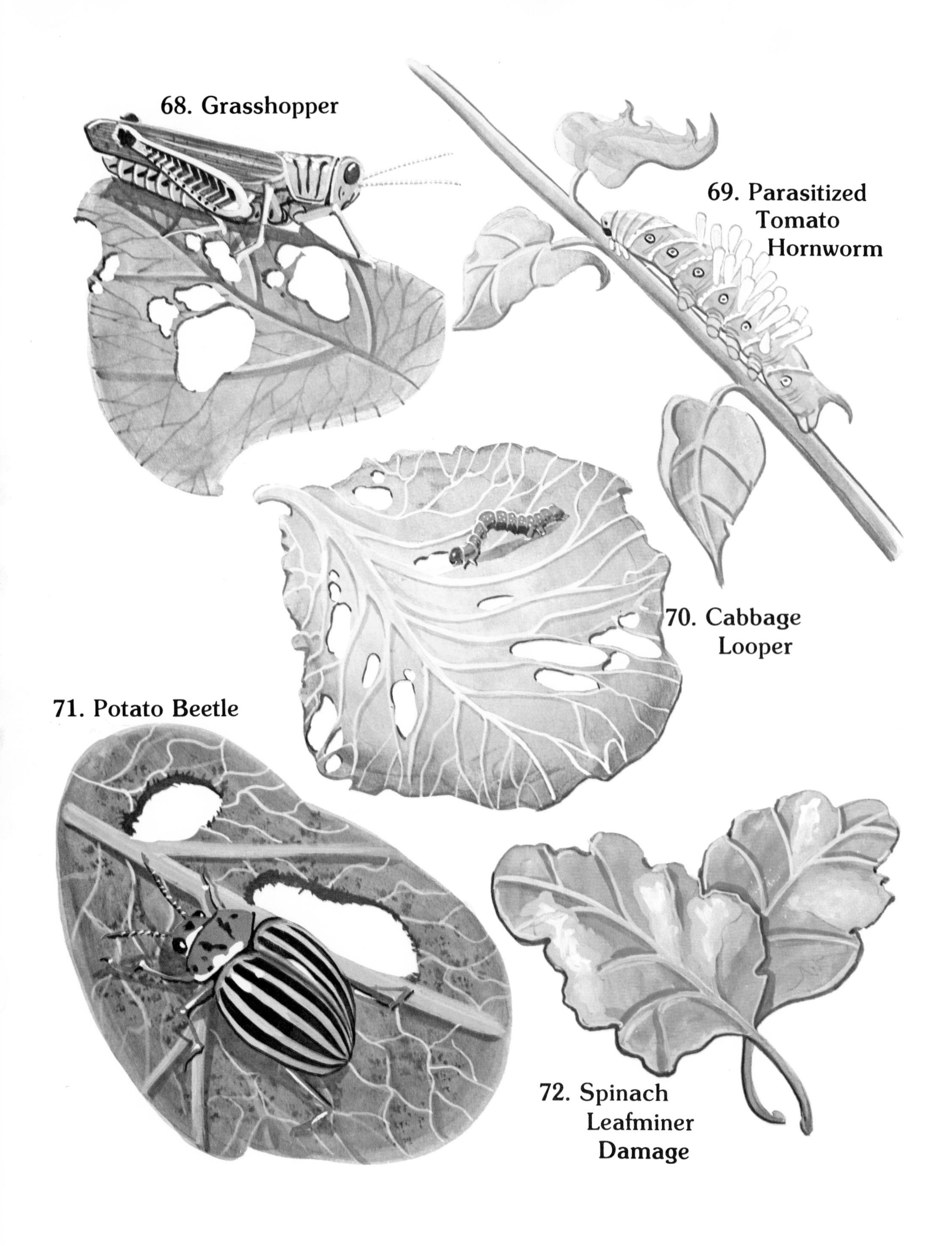
68. Grasshopper
69. Parasitized
Tomato
Hornworm
70. Cabbage
Looper
71. Potato Beetle
72. Spinach
Leafminer
Damage

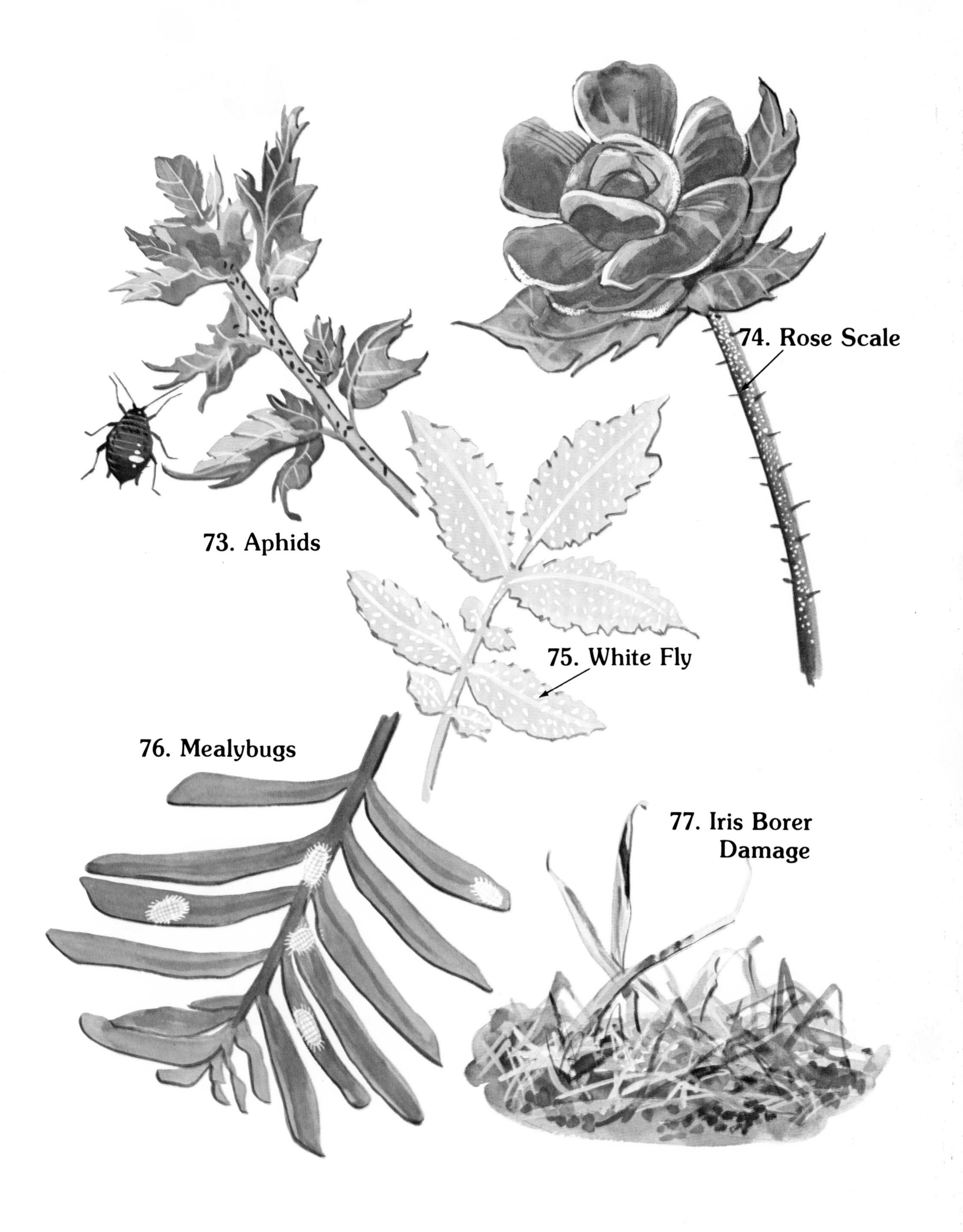
73. Aphids
74. Rose Scale
75. White Fly
76. Mealybugs
77. Iris Borer
Damage

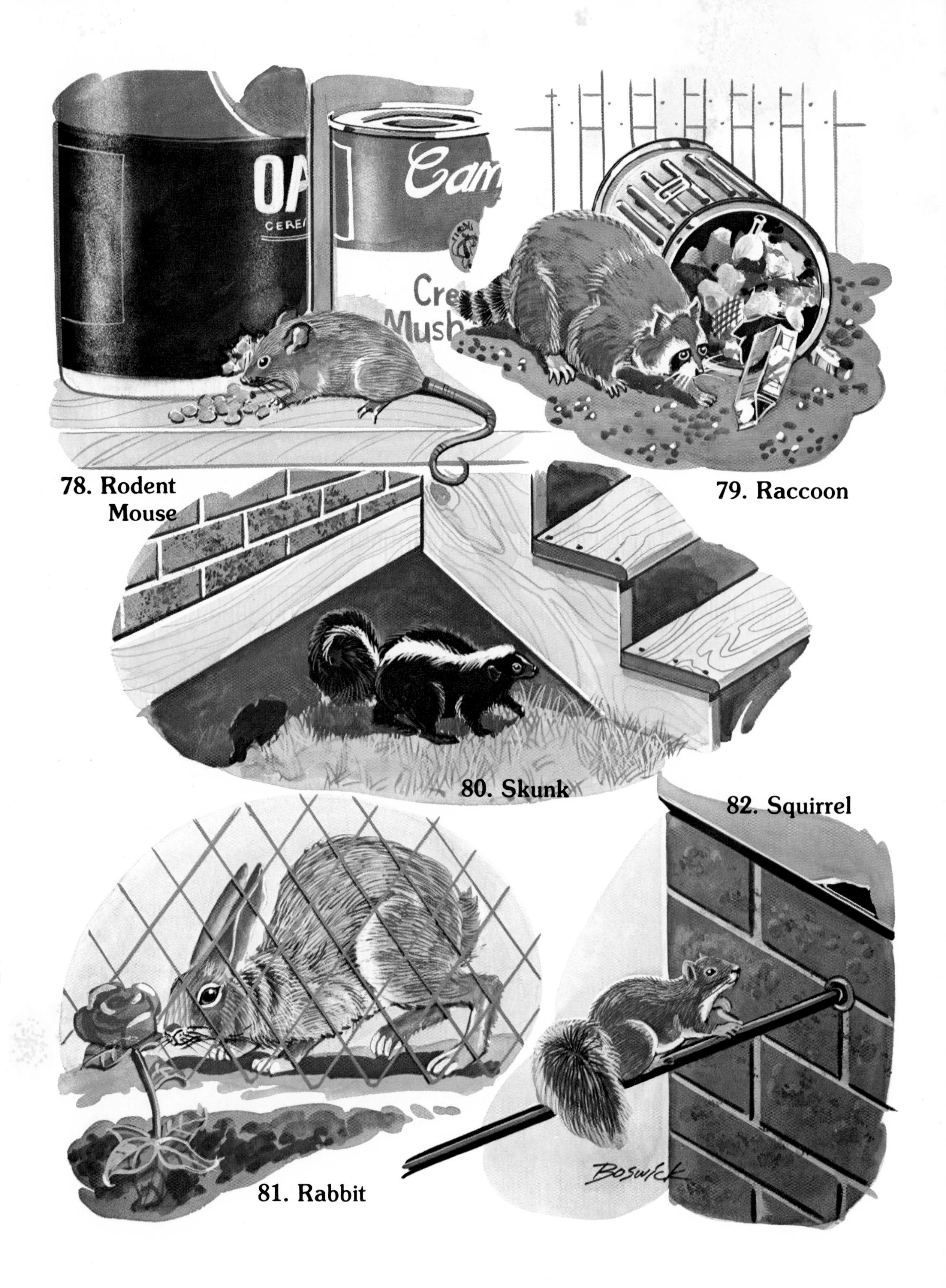

78. Rodent Mouse

79. Raccoon

80. Skunk

81. Rabbit

82. Squirrel

Sowbugs (see, "Lawn Pests")

Spiders

"Little Miss Muffet sat on her tuffet, eating her curds and whey. Along came a spider and sat down beside her, and frightened Miss Muffet away."

Despite the enormous popularity of *Charlotte's Web,* almost everyone is afraid of spiders. The very word "spider" conjures up images of tarantulas and black widows. Actually, most spiders are not dangerous except to individuals who are hypersensitive to their venom. Although all spiders are venomous (that's how they kill their prey), they feed on insects, small animals and other spiders, and rarely bite people.

Two species that are dangerous to man are the brown recluse and the black widow. If you live in an area where these spiders are common, special precautions are in order.

Black Widow Spider

Black Widow

Although the black widow has venom 15 times more powerful than that of a six-foot rattlesnake, such a small amount is injected that death is extremely rare. Only the female spider bites (Color Figure 32). You can identify her by her shiny black body and legs and the red hourglass pattern on her abdomen.

Black widow venom is a neurotoxin that affects the central nervous system. The bite feels like a pinprick and is followed by a stinging or burning sensation. Reactions include nausea, trembling, fever, abdominal pain and cramping, depression, and speech difficulties. The victim should be calmed and taken to the nearest hospital for treatment. Respiratory paralysis leading to death may develop in children.

Brown Recluse

In general, the brown recluse is as timid as its name implies (Color Figure 35). It lives outdoors in warm climates but needs shelter to survive in colder regions. You will find the brown recluse in out-of-the-way spots like attics, closets, and garages. It attacks only when disturbed.

The victim usually feels only a slight stinging sensation when bitten, but blistering, swelling, and severe local pain soon develop. The tissues affected by the venom die and gradually slough off, leaving an ulcerated wound. The final result is a sunken scar. In addition to these local reactions, the victim often experiences chills, nausea, fever, and a generalized rash.

Early diagnosis and treatment are essential because untreated bites can be fatal. Surgical excision of the affected tissue is the best method, but the excision must be done within a few hours, before the venom spreads. The incision and suction method

Brown Recluse Spider

used for snakebite is ineffective because brown recluse venom travels by gravity rather than through the lymphatic system as snake venom does. For this reason, the bitten area should not be elevated to alleviate pain. When excision is impossible, corticosteroids and antihistamines seem to reduce systemic reactions.

Control

Pesticide sprays provide excellent control. Treat the foundation of your home and a four-inch strip of soil around the sides (do not spray flowers and shrubs). Use at least four gallons of diluted spray for thorough coverage. Drench until the spray begins to drip off. You should also treat a two-inch strip around the patio and along sidewalks.

Indoors, treat baseboards and crevices. Foundation spraying every six weeks during the warmer months will make additional indoor treatments unnecessary.

Springtails

Springtails are unattractive but harmless insects that live primarily on damp and decaying organic matter. Their name derives from the fact that they tuck their tails under their bodies and then propel themselves forward by releasing their tails. Although they can be found outdoors in enormous numbers, springtails enter houses more or less by accident, usually through an open window or broken screen. They do no damage inside, but it can be quite startling to find one in the kitchen or bathroom. People with houseplants are more apt to have springtail problems.

Good sanitation generally provides sufficient control. Clean out all decaying vegetation near the foundation of your house. If they do wander indoors, space spraying does an excellent job. Treat window sills, door frames, and baseboards with special care. You can use your vacuum cleaner for picking up springtails.

Termites

Termites, also known as white ants, are social insects that feed on wood products. About 2000 species have been identified, but most are limited to the tropics. The enormous mounds they construct (up to 40 feet high!) are common sights in Australia and Africa.

Subterranean Termites

A colony usually consists of three castes—a royal couple (a sexually mature male and female), workers, and soldiers who defend the others in times of danger (Color Figures 21 to 25). Because colonies require a constant supply of moisture, they are located underground.

People Pests

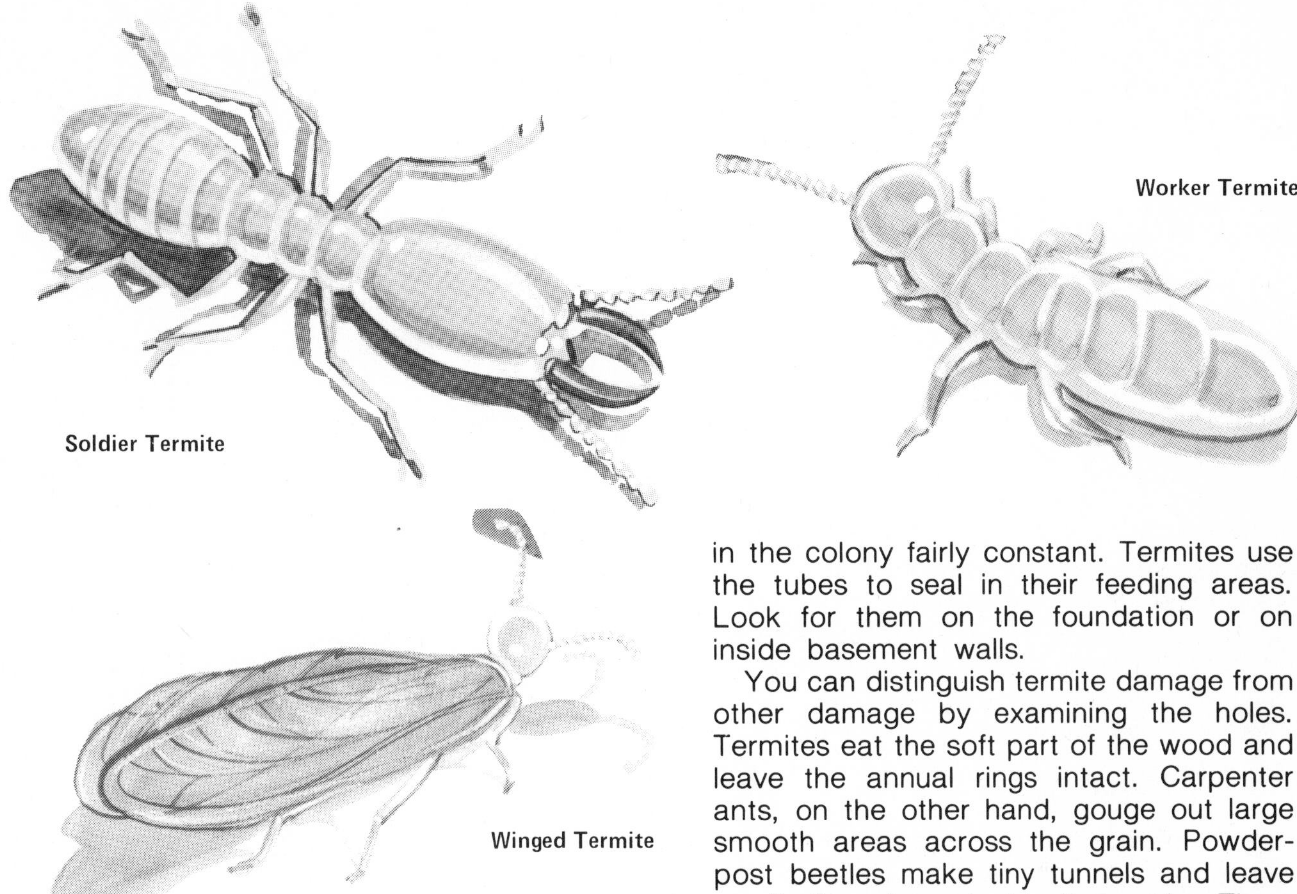

Soldier Termite

Worker Termite

Winged Termite

Despite their reputation, termites are not always harmful. They may live for years in the backyard, peacefully eating an old tree stump, and never bother your house at all. Even after they move inside, it may be eight to ten years before any structural damage is done.

Termites can find a surprising number of ways into a building. Any wood in contact with soil (basement windows, porches, siding, etc.) is perfect for them. If no wood is present, they can use cracks in the foundation or any other small openings into the house. Even brick houses may be infested.

If you suspect that your house has been invaded, look for mud tubes. These tubes serve two purposes. They provide pathways over obstructions termites cannot chew through and they keep the humidity in the colony fairly constant. Termites use the tubes to seal in their feeding areas. Look for them on the foundation or on inside basement walls.

You can distinguish termite damage from other damage by examining the holes. Termites eat the soft part of the wood and leave the annual rings intact. Carpenter ants, on the other hand, gouge out large smooth areas across the grain. Powder-post beetles make tiny tunnels and leave small piles of sawdust in their wake. There are also several fungi that cause wood to

Mud Tubes

IS IT AN ANT OR A TERMITE?

Characteristics	Ant	Termite
Wings	Two pairs, with the front pair much longer than the back pair	Two pairs of equal length
Antenna	Bends at right angles	Straight
Abdomen	Wasp waist (pinched in)	No wasp waist

rot, but the wood will appear charred or crumbly and there will be no tunnels.

Termites are one problem you cannot handle yourself. Call in a professional as soon as possible.

Ticks

Some ticks are dangerous and they are all real nuisances. This is one problem that never should be ignored!

If you enjoy outdoor pleasures such as camping, hiking, fishing, berry-picking, or even picnicking, be especially cautious in June when tick populations are at their peak. Ticks cling to vegetation in woody or shrubbed areas, waiting for an unwary warm-blooded animal to pass by. One of their favorite feeding spots is the area of the head around the hairline (Color Figure 13).

Ticks

Since they are capable of carrying diseases such as Rocky Mountain spotted fever, it pays to take precautions. Use a tick repellent to cover clothing and exposed parts of your body. An effective repellent will usually keep mosquitoes away as well. Depending on how active you are, four or five applications per day may be necessary. If you plan to spend several hours in one spot, consider spraying nearby grass and shrubs.

Ticks are unlikely to be an indoor problem unless you have a dog. Most dog-owners, however, will face an infestation at least once if they don't take precautions. Your pet will unknowingly carry them inside. There are two major tick species to watch out for.

American Dog Ticks—These little creatures usually live on mice or other wild animals. If you have mice in the walls, hope that they are tick-free. The newly hatched ticks, called seed ticks, feed on small rodents and then drop to the ground where they molt and become adults. The adult ticks, also known as wood ticks, attack larger animals like dogs and humans. They are very active in the early spring.

Brown Dog Ticks—If you find a tick or two crawling on the wall or on furniture, they are apt to be brown dog ticks. Although they do not bite humans, they will make your pet's life miserable. Surprisingly, the real problem develops in late winter rather than the fall. This is because the female tick that your dog carried inside in autumn lays eggs from December to February. Once the eggs hatch, you won't know whether to call in an exterminator or a psychiatrist to help you cope.

Control

Controlling ticks is relatively simple. Flea and tick collars or dusting powders will protect your pet. The same powder can also be used on floors, rugs, and furniture. Vacuuming removes it once the job is done. Spray baseboards and crevices thoroughly if the problem is serious. Pressurized aerosol foggers can also be used as a supplement if you are desperate.

Control can be achieved with a little effort. Clean air registers regularly to prevent dust and lint accumulations. Dry-cleaning clothes will kill these pests. Clean woolens should be stored in plastic bags inside tightly closed chests. If necessary, spray storage and infested areas.

Tissue Paper Beetles

Tissue paper beetles are rapidly becoming a major annoyance. Perhaps the most irritating thing about them is that they like clean clothes and clean bedding, rather than soiled ones. Thus, in most instances it is the fastidious housekeeper, tidying drawers and closets, who encounters them. Occasionally, they live under the eaves in the cocoons of other insects, making a quick trip to the interior of the house easy. You may find them in china closets or bookcases as well (Color Figure 6). Although the life cycle of a tissue paper beetle usually takes one year, the larvae have been known to survive three or four years without food.

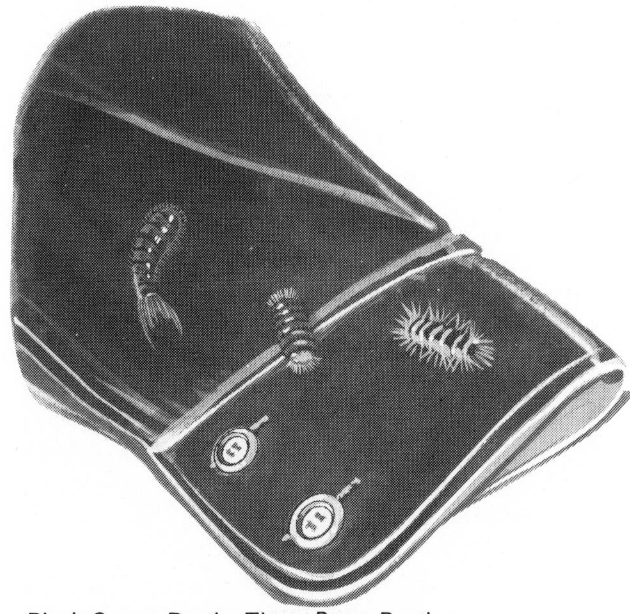

Black Carpet Beetle, Tissue Paper Beetle, and Carpet Beetle Larvae

Wasps

Wasps, like bees, have nasty reputations that are not entirely deserved. Most wasps prey on other insects or small creatures such as spiders and centipedes. Because they feed on plant-eaters that compete with man for food, wasps on the whole are beneficial.

There are about 20,000 wasp species in the world, ranging in size from less than one-eighth inch to more than three inches long. Most species are solitary but a few, including the most familiar ones, are social and live in colonies. Scientists classify insects as social if: (1) individuals cooperate in caring for the young; (2) sterile individuals provide food or care for individuals capable of reproduction; and (3) two or more overlapping generations contribute to the work force (that is, offspring assist parents with the labor needed to sustain the colony).

Many people regard wasps as dangerous because they sting. This fear is generally unjustified. Perhaps 90% of the wasp species are solitary and the majority have stings that are nontoxic to humans. In addition, many wasps are so small that they are incapable of piercing the relatively thick skin of man. A few species, however, can administer painful stings. Usually, these result only in local pain and swelling that soon subside. People who are allergic to wasp venom may experience a generalized rash, an intense burning sensation on the skin, respiratory difficulties, fainting, and, in extreme cases, an extended period of semiconsciousness. Hypersensitivity may develop after one sting, so those who

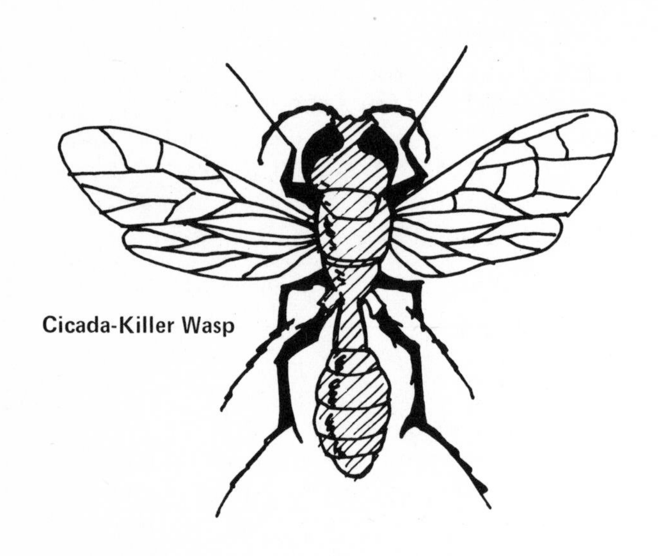

Cicada-Killer Wasp

have been stung previously should be extra-careful when dealing with wasps.

Cicada-Killer Wasps

The cicada-killer is one of the largest wasps in existence. It is black with broad yellow bands and is often mistaken for a yellow jacket. The cicada-killer is a ground-nester that preys on cicadas or locusts and thus helps the gardener. Its sting paralyzes the cicada, which serves as a living host for the wasp egg. After the egg hatches, the young wasp feeds on the cicada.

Cicada-killers are not aggressive towards people but will sting when disturbed. Avoid provoking them because the results can be quite unpleasant!

Paper Wasps

Paper wasp nests can be seen hanging from tree limbs or the eaves of a building. Look for several umbrella-shaped nests with black and yellow wasps resting on them. Paper wasps, like cicada-killers, are often beneficial because they prey on caterpillars. They will become angry and aggressive if you disturb their nests.

Parasitic Wasps

These harmless creatures are often mistaken for ants. You can distinguish them from ants by the shape of their antennas. The antennas of ants bend at right angles, while those of parasitic wasps have a graceful curve.

You may find various kinds of parasitic wasps inside the house, but there is no need for alarm. They do not attack people or damage anything you own. In fact, they eat caterpillars and other insect pests, and are most numerous in the summer months when these insects hit their population peak.

Yellow Jackets

Yellow jackets usually appear as soon as warm weather begins and are among the last to disappear in the fall. They are social insects and, depending on the species, build paper-like nests above or below the ground. Most are black and yellow and a little more than one-half inch long. One exception is the bald-faced hornet, common in the Midwest, which is large and black. (Color Figure 28).

A typical colony contains an egg-laying queen, several thousand sterile female workers, and fertile males that appear later

Wasps' Nest

in the season. The workers construct cells for the eggs and feed tiny bits of meat to the larvae. The larvae, in turn, feed the adults with small, clear droplets of partially digested food. This unusual interdependence creates a strong bond. The adults also feed on nectar and other sweets (they love soft drinks and cotton candy and can be real nuisances at outdoor carnivals).

Look for aerial nests in trees, shrubs, buildings, and even machinery. The nests, which are sometimes as large as basketballs, are oval-shaped and gray. (Color Figure 30). They are constructed of bits of bark and wood that have been chewed into a coarse paper.

Control

Controlling wasps is fairly simple but a bit risky. Don't attempt to do the job yourself if you have been stung in the past or if you might be allergic.

Insecticides applied directly to the nests are very effective. Aerial nests can sometimes be covered with a heavy plastic bag such as a trash can liner. Insert an impregnated resin strip before closing the bag. This will kill the wasps. A local biology teacher might like the nest for his class to study. To eliminate underground nests, apply at least one gallon of diluted insecticide to the nesting site, soaking the soil thoroughly, and then cover it over. All control measures should be taken at night when the workers have returned to the nest.

Weevils

Weevils are a branch of the beetle family distinguished by their long snout-like noses. They are primarily outdoor pests, feeding on a wide variety of plants, but often migrate into the house when populations become heavy. The three types of weevils you are most likely to encounter indoors are the nocturnal black vine weevil, the clover leaf weevil, and the strawberry

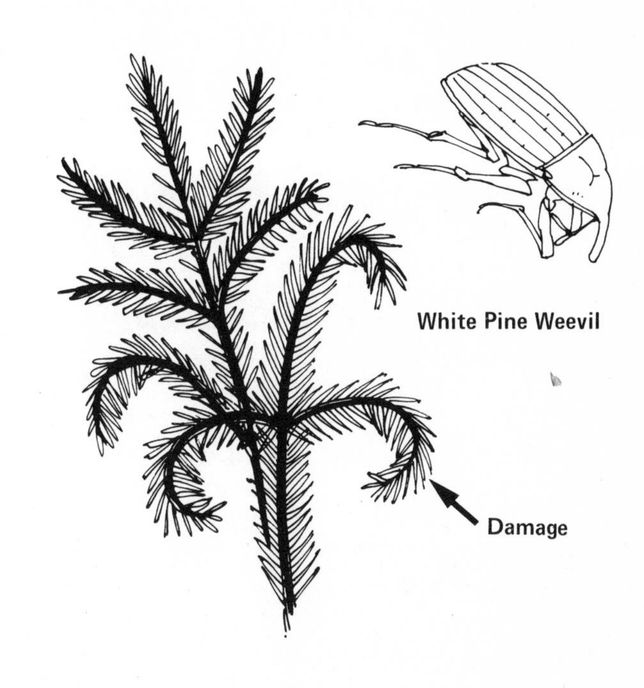

root weevil. None of them do much damage indoors (although they are real nuisances in the garden), but they make unattractive houseguests.

Foundation spraying, including a strip of soil around the house, is usually sufficient to discourage them from entering. If you have already spotted them crawling around the house, treat baseboards, floor and wall crevices, and thresholds.

Wharf Borers

These nautical-sounding beetles can be found inland as well as in coastal areas. They like moist wood and are particularly fond of timber that has been in contact with saltwater. Replace rotted and infested wood as soon as possible. Spraying with a wood preservative is helpful.

Houseplant Problems

Houseplants can brighten a room while satisfying the ambition of a gardener who prefers to stay indoors. Nonetheless, they can attract some annoying pests who might not bother you otherwise.

Most of the insects that the backyard gardener encounters pose no threat to houseplants, but the few who do can devour your favorites fast. In addition, they may leave the plants in order to explore the rest of house, becoming real nuisances in the process.

Aphids, mealybugs, white flies, springtails, mites, fungus gnats, and some scale insects are the most likely culprits. They may enter the house with a new plant or fly in through an open door or window. Any of the pests on your trees and shrubs may wander inside and find a lovely surprise in the form of your most cherished fern or African violet.

Prevention is better than cure. You can keep problems to a minimum by taking a few simple precautions. Houseplants should be kept inside. If they must be moved outdoors, keep any exposed plants isolated from the others for three or four weeks after bringing them back in. A similar quarantine should be observed when you purchase a new plant. If one plant does become infested, the pests are less likely to spread to the rest.

Using insecticides on houseplants is tricky. Some of these plants are very delicate and the insecticide may be just as damaging as insects. If you feel that you must use one, follow label directions carefully. Overapplication can be toxic or even deadly to your plants.

Before taking drastic measures, be sure the plants are actually infested. Insects and mites are often blamed for damage caused by other factors. If you notice any symptoms, try correcting the environmental causes listed in the chart before you panic.

DO YOU REALLY HAVE A BUG PROBLEM?

Symptom	Possible Environmental Cause
Sudden leaf drop	Cold air, transplant shock
Spotted leaves	Sunburn, water spill on leaves
Yellowing of lower leaves with tip browning on upper leaves	Improper watering, wrong temperature, pot-bound plant.
Scalded or browning leaves or tips	Too much light, too little water; drafts; root damage
Yellowing and dropping of lower leaves	Pot too small; improper fertilization; cold air; improper watering; lack of drainage; root damage; too little light

Once you are sure that you have a bug problem, try wiping the stems and leaves with a mild soap-and-water solution. Rinse the plants thoroughly in the sink or bathtub afterwards. It's best to do this at your normal watering time. If you are the patient type, you can eliminate some small colonies by dabbing each insect with a cotton swab dipped in alcohol.

Many problems can be solved by switching plants. Some houseplants are much more resistant to insects than others. Even if you are devoted to some of the more finicky ones, however, these suggestions should help.

SUGGESTIONS FOR CONTROLLING PEOPLE PESTS

Pest	Control	Comments
Ants	Chlorpyrifos, Diazinon, propoxur	Indoors — treat baseboards, wall and floor crevices, and thresholds; outdoors — spray foundation and a four-inch strip of soil around the house

People Pests

Pest	Control	Comments
Aquatic insects	Malathion, pyrethrin	Treat grassy areas near pool; do not spray pool itself; use aerosols for temporary control
Assassin bugs	Chlorpyrifos, Diazinon, propoxur	Destroy wild animal dens; spray foundation and surrounding soil
Backswimmer	————————	See Aquatic insects
Batbugs	Malathion, pyrethrin	See "Small Animal Pests" for bat control; treat bedframe, slats, and springs thoroughly; dust seams and tufts on mattresses
Bedbugs	Malathion, pyrethrin	Treat bedframe, slats, and springs thoroughly; dust seams and tufts on mattresses
Bees	Carbaryl, impregnated resin strips	Hang strips in the attic to prevent nesting; wall nests — apply insecticide through hole and seal; remove nest within two weeks
Black vine weevils	————————	See Weevils
Bluebottle flies	Impregnated resin strips, malathion, pyrethrin	Eliminate standing water; spray grass and shrubs near doorways and trash cans; place strips in trash cans; use aerosols indoors
Booklice	Pyrethrin	Keep books and papers dry; improve ventilation; spray shelves and infested areas
Boxelder bugs	Carbaryl, Diazinon, pyrethrin	Indoors — vacuum; outdoors — spray infested areas, eaves, and foundation
Carpet beetle	Chlorpyrifos, Diazinon, propoxur, pyrethrin	Wash or dry-clean fabrics; spray baseboards and storage areas
Centipedes	Carbaryl, Diazinon	Indoors — vacuum; outdoors — spray foundation and grass within ten feet of the house
Cereal insects	Chlorpyrifos, Diazinon, propoxur, pyrethrin	Discard infested packages; vacuum shelves; if necessary spray crevices in cabinets or shelves
Chiggers	————————	See Mites
Clothes moths	Chlorpyrifos, Diazinon	Wash or dry-clean fabrics; keep air-shafts clean; spray storage areas, baseboards, and carpet backing
Cluster flies	Impregnated resin strips, malathion, pyrethrin	Indoors — hang strips out of reach of children and away from food-handling areas; outdoors — spray walls and eaves in the fall and seal all cracks
Cockroaches	Chlorpyrifos, Diazinon, Propoxur	Treat infested areas and possible hiding places — under sink and other appliances, baseboards, and crevices; repeat in 30 days if necessary

People Pests

Pest	Control	Comments
Crickets	Chlorpyrifos, Diazinon, propoxur	Indoors — treat baseboards, wall and floor crevices, and thresholds; outdoors — spray foundation and a four-inch strip of soil around the house
Daddy longlegs	————————	No control needed
Deer and horse flies	Deet	Only personal protection is necessary
Drainflies	Malathion, pyrethrin	Indoors — clean out all drains; pour hot water or rubbing alcohol into overflow drains to kill maggots; spray only as a last resort; outdoors — spray shrubs, tall grass, and area around trash cans
Dry rot	Penta	Replace decaying wood; store lumber above the ground and away from damp; use a wood preservative
Earwigs	————————	See "Lawn Pests"
Elm leaf beetles	Carbaryl, pyrethrin	Indoors — vacuum; spray only if necessary; outdoors — see Tree and Shrub pests
Fleas	Carbaryl, Diazinon, pyrethrin	Dust pets (do not use on kittens less than four weeks old) and their favorite resting spots; treat rugs and upholstered furniture, let stand 24 hours, and then vacuum
Fruit flies	Pyrethrin	Thoroughly clean breeding media; drain standing water; spray only as a last resort
Giant water bugs	————————	See Aquatic insects
Gnats	Impregnated resin strips, malathion, pyrethrin	Indoors — use fly swatter or space sprays; outdoors — place strips in garbage cans; eliminate standing water; spray shrubs and tall grass near doors and in garbage can area
Houseflies	Impregnated resin strips	Indoors — hang strips out of the reach of children and away from food-handling areas; outdoors — clean up animal droppings immediately
Larder beetles	Chlorpyrifos, Diazinon propoxur	Treat baseboards, wall and floor crevices; look for dead animal body
Lice	Carbaryl, malathion	Dust body hair lightly, avoiding eyes; wash clothing and bedding; check with your physician
Millipedes	————————	See Centipedes
Mites	Deet, dicofol, pyrethrin	Consult a dermatologist; take a warm, soapy bath or shower after exposure; use a repellent; indoors — vacuum; outdoors — spray walls and grassy areas within ten feet of the house; replant soil along foundation with plants that do not attract clover mites

People Pests

Pest	Control	Comments
Mosquitoes	Impregnated resin strips, malathion, pyrethrin	Indoors — use fly swatter, strips, a space spray; outdoors — place strips in garbage cans; spray shrubs and tall grass near doors and in garbage can area; drain standing water
Picnic beetles	Carbaryl, Diazinon, malathion	Pick crops before they become overripe; spray grassy areas and near garbage cans; repeat in four or five days if necessary
Pillbugs	————————	See "Lawn Pests"
Powder-post beetles	Lindane, penta	Saturate infested unfinished wood; use penta as a preservative
Scorpions	Chlorpyrifos, Diazinon, propoxur	Remove brick and lumber piles; spray foundation, wall crevices, and other hiding places
Silverfish	Chlorpyrifos, Diazinon, propoxur	Store books and papers away from damp; treat baseboards, closets, and pipe openings in the wall; repeat in 14 days if necessary
Sowbugs	————————	See "Lawn Pests"
Spiders	Chlorpyrifos, Diazinon, propoxur	Indoors — treat baseboards, crevices, and thresholds; outdoors — spray foundation and a four-inch strip of soil around the house; repeat every six weeks during the warmer months
Springtails	Diazinon, malathion	Indoors — vacuum; outdoors — fill in moist spots around foundation
Termites	————————	Call in a professional exterminator
Ticks	Carbaryl, Diazinon, pyrethrin	Dust pets (do not use on kittens less than four weeks old) and their favorite resting spots; treat rugs and upholstered furniture, let stand 24 hours, and then vacuum
Tissue paper beetles	Chlorpyrifos, Diazinon	Dry-clean clothes; clean out air registers; if necessary spray storage and infested areas
Wasps	Carbaryl, impregnated resin strips	Indoors — hang strips in the attic to discourage nesting; outdoors — spray aerial nests at night; drench soil around ground nests and cover over
Water boatman	————————	See Aquatic insects
Weevils	Chlorpyrifos, Diazinon, propoxur	Indoors — treat baseboards, wall and floor crevices, and thresholds; outdoors — spray foundation and a four-inch strip of soil around the house
Wharf borer	Lindane, penta	Replace rotten and infested wood; use penta as a preservative

Tree and Shrub Pests

No matter what kind of tree or shrub you have, there is some insect specially adapted to thrive on it. Don't let these pests destroy your beautiful — and valuable — plantings. Be alert for the symptoms that indicate the presence of pests and take action fast.

Insecto
Cider

Tree and Shrub Pests

WHAT DOES a tree mean to you? Do you ever think of it as a woody perennial plant with one main stem and many branches? Probably not! Most people have a favorite tree, one they love for its beauty or for its association with happy memories. Does seeing a tree take you back to your first nature lesson, to a treehouse, or to the rubber-tire swing that delighted you as a child? Do you think of picnics under the green leaves of summer or the gaily decorated Christmas tree of winter?

Obviously, a tree means something different to each individual. All of us, however, should be aware that trees are subject to the ravages of insects and disease. In addition to natural pests, trees suffer from ecological destruction caused by urbanization. This is particularly true of oaks. A drive or stroll through a wooded neighborhood will show you oak trees in various stages of decline. The proof is easily seen, ranging from a few bare tips to large dead limbs.

Construction is the culprit. When forested areas are developed, the harmonious ecological system is destroyed. The contractor removes trees, clears away shrubs, rakes organic decaying material that serves as a natural fertilizer, compacts the soil with heavy equipment, changes the drainage system by constructing trenches and pipelines, paves over the soil, and lays down sod that competes for surface moisture and nutrients. The oak is especially vulnerable because of its shallow root system.

Many people have paid extra for wooded lots, only to find that the trees die within a few years, and that the beauty and value of their property have suffered as a result.

There are measures you can take to prevent or at least minimize this destruction. Although there is little or nothing that an individual can do to stop urbanization, proper tree care can give your trees a better chance to survive by keeping them as strong and healthy as possible. A tree weakened by pests and disease is more likely to succumb to environmental change.

Perhaps the best method of discussing common tree and shrub problems is to list easily recognizable symptoms and suggest possible causes and solutions.

The Symptoms To Watch For

An alert homeowner can often spot signs of damage at an early stage when effective action can still be taken. Make a habit of examining your trees and shrubs at frequent intervals. Start checking in the early spring before new growth begins.

Holes In The Leaves—There are many insects that eat holes in leaves. You should be able to spot the insects during the day or early evening. Most leaf-eating pests can be controlled by spraying. Freezing weather or hailstorms may also cause holes to appear.

Spots On The Leaves—This symptom can be caused by a variety of problems. Among the most likely are spray burn, disease organisms, and insect galls or excrement. Some plant-sucking mites and insects cause spots to form by removing chlorophyll from the leaves. Insecticides can control this problem.

Leaf Browning—If leaves are brown along the edges, the cause may be chemical or fertilizer burns, lack of water and nutrients, transplant shock, too much water (which can kill roots), or leaf-mining insects. These insects can be controlled with pesticides.

Leaf Yellowing—This symptom may be the result of nutritional deficiency such as lack of iron, roots injured by digging, girdling, or too much dirt fill, too much water, or sucking insects or shade tree borers. Controlling borers is possible, but the species must be identified before any particular treatment can be recommended.

Leaves Curled Or Distorted—This can be

caused by freeze damage to a vein or leaf stalk (petiole), weed killer, a heavy population of insects such as aphids, leafrollers, or insect galls which cut off the nutrient supply. Insects can be controlled by pesticides. Generally speaking, it is not necessary to control insect galls.

Green Leaves Dropping—This can be the result of freeze damage or an overabundance of a substance such as fertilizer, herbicide, or plant hormones. Hormone imbalances are very difficult to diagnose, but they play an important role in this condition. When the tree is a maple, the damage may also result from the activities of maple petiole borers, pests that can be controlled with insecticide.

Live Branches Dropping—If the branch is broken, check for squirrels; if it is sawn, check with your neighbors. If it is definitely cut, look for such insects as the twig pruner or twig girdler. No spraying is needed at this point; just collect and destroy the fallen branches. Consider it your exercise for the day.

Branch Dying—This symptom has a number of possible causes: roots injured by girdling, digging, machines, too much dirt fill, a gas leak, toxic substances or too much fertilizer in the ground, excessive heat, or insect borers.

The Pests

If your tree is infested, you must first identify the problem (either with the help of this book or by calling in a qualified tree specialist) and then apply the proper control measures. Finally, prune dead or dying branches and water if necessary. Then you've done all you can—the rest is up to the tree.

This section deals with nonspecific pests—those that attack a wide variety of trees and shrubs. Some—such as the Japanese beetle—are species that are happy to eat almost anything. Others—such as tree borers—are pest categories that include many different species. Individual species within the category may concentrate on certain plants.

Aphid

Aphids

Aphids are plant-sucking insects commonly called plant lice. There is probably a species of aphid for virtually every species of plant. Aphids are usually greenish in color, but they may be pink, black, or brown. They are fragile, awkward-looking creatures, somewhat pear-like in shape. Look for two horn-like protuberances on their rear ends and strong, slender beaks. Once these beaks are inserted into the plant, aphids can suck out large amounts of the plant juices. Aphids are the Draculas of the insect world.

Much of the damage they cause is due to the fact that they reproduce rapidly and congregate in great numbers. It has been estimated that one aphid can produce 1521^{21} offspring in a single season. The sweet-smelling, sticky substance they secrete, called "honeydew," drips down from the tree and covers everything left underneath it. It also provides a growth medium for a black sooty mold that can interfere with photosynthesis.

Fortunately, aphids have many natural

enemies that prey on them and help keep the populations down to a tolerable level.

Brown-Tail Moths

During its caterpillar stage, this moth feeds on the foliage of fruit and shade trees. Among its favorites are apple, maple, pear, plum, and willow trees.

The adult moth, which is white with brown hairs on the abdomen, lays its eggs on the undersides of the leaves. The reddish-brown caterpillar spends the winter in webbed areas of leaves. It feeds for a short time in the spring before forming a cocoon. The adult emerges in July and mates immediately.

Although no effective method of control has been discovered as yet, the brown-tail moth is a real pest only in the New England states.

Cankerworms

For more than two centuries outbreaks of cankerworms, also known as spanworms, measuring worms, or inchworms, have periodically defoliated shade and fruit trees in various sections of the United States. In unsprayed or poorly sprayed orchards, they may cause complete defoliation and loss of the crop, but they are of no importance in well-sprayed orchards.

The foliage of the trees is eaten and skeletonized by spring cankerworms. The injury occurs just about the time the trees have come into full foliage. Silken threads are spun from branch to branch on the tree and from the branches to the ground.

Brown and brownish-green measuring worms about one inch long spin down from the tree when it is jarred or shaken. Heavily infested orchards look like they have been scorched by fire.

The cankerworm passes the winter in the form of a naked brown pupa about one-half inch long by one-eighth inch thick. These pupae are found in the soil from one to four inches below the surface, and in greatest numbers close to the base of the trees.

Inchworm Larva

The moths begin emerging during warm periods in February and continue coming out until the end of April. The male moth is strongly winged and is dull gray, much the color of a well-weathered piece of board. These moths may be seen flitting from tree to tree at dusk and after dark on spring evenings.

The female moth is wingless, with a gray spider body. She differs from the fall cankerworm female by having a dark stripe down the middle of the back and two transverse rows of small reddish spines across each abdominal segment on the upper side of her body.

On emerging from the ground, the female crawls to a tree and up the trunk, or onto the branches, where she mates with the male and deposits her oval dark-brown eggs in irregular masses under the loose scales of bark. These eggs hatch about one month later into small greenish or brownish measuring worms, which at once begin to feed on the foliage. These worms are distinguished from the fall cankerworm by having only two pairs of prolegs, near the end of the body. They vary from light-brown to nearly black and usually have a yellowish stripe below the spiracles. The underparts are partially black.

When not feeding, the larvae (Color Figure 51) tend to rest upon the twigs more than upon the leaves. They feed for three weeks to one month and, if abundant, may completely strip the foliage from the trees. At the end of the feeding period, they crawl or spin down to the ground where they excavate the small cells in which they change to the pupal stage and pass the remainder of the summer and the following winter.

Insecticides will control cankerworms easily. *Bacillus thuringiensis,* a disease organism, is applied as a spray and is very effective.

Treatment of large trees with insecticides or *Bacillus thuringiensis* is really not necessary because these large deciduous trees will leaf out three or four times a year, so those trees that are defoliated for a couple of weeks will soon have new leaves. Spraying infested smaller shrubs and young saplings is probably a good idea.

A band of sticky material such as Tanglefoot around the trunk of the tree, from two to four feet off the ground, may help to control this worm. However, this is impractical in a heavily wooded area.

Gypsy Moths

Like its famous namesake Gypsy Rose Lee, this insect is a terrific stripper! One caterpillar can eat a square foot of leaf surface every 24 hours.

The Gypsy moth was introduced to the United States in 1869 by a naturalist who wanted to study them. A violent windstorm accidentally freed a number of moths and they began to spread rapidly because the natural predators and parasites that keep them under control in Europe are not found in this country. They often use the most modern methods of transportation, hitching rides on campers, automobiles, and railroad cars. The egg masses deposited on these vehicles are seldom noticed, so the infestation spreads to new areas.

The caterpillar grows to be about two inches long and has a conspicuous double row of blue and red dots down its back. It is during this stage of its existence that damage to the plants occurs. A heavy population can defoliate a forest.

Japanese Beetle

The adult moth does not feed. The males have dark-brown forewings and wingspans of about 1½ inches and are strong fliers. The females are white with wings marked with black. They are much larger than the males but do not fly.

Japanese Beetles

Population explosions among Japanese beetles mean destruction to a great variety of trees, shrubs, and other leafy plants.

First seen in the United States in 1916, this native of the Orient has spread across the country. In recent years, government agencies have instituted strict measures in an attempt to check their dispersion.

The adult beetles are easy to recognize because of their brilliant coloring. Their bodies are a metallic green with white spots on the abdomens and their wings are copper. They are strong fliers.

Eggs are deposited in the soil, and the newly hatched grubs feed on the roots of grasses. They live below the frostline in the soil during the winter and emerge as adults in June and July. The adults feed on apples, cornsilk, grapes, and foliage of all sorts for about three months.

Insecticides can provide a good deal of protection.

Leafhoppers

Leafhoppers can be found on almost every

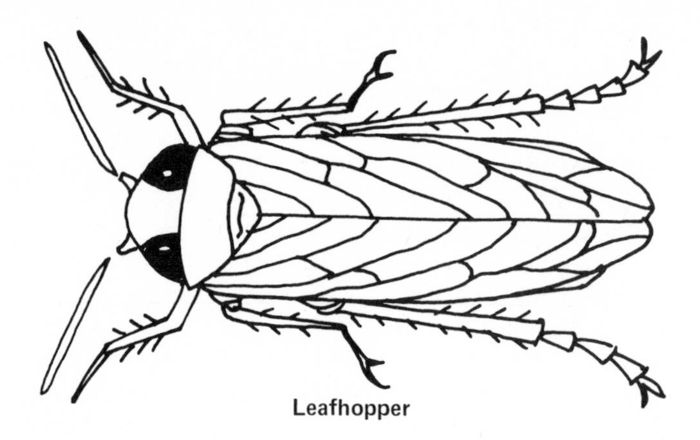

Leafhopper

type of plant, including forest, shade and orchard trees, shrubs, grasses, and flowers, and many field and garden crops. Most species produce a single generation of offspring in a year, but a few have two or three generations in a year. The winter is passed either in the adult or the egg stages, depending on the species.

There are many economically important pest species in this group. They are the cause of five major types of plant damage:

1. Some species on apple trees remove great amounts of sap and reduce the chlorophyll in the leaves, which then become covered with minute white or yellowish spots. Excessive feeding produces yellow or brown leaves.
2. The potato leafhopper interferes with the normal physiology of the plant, plugging vessels in the leaves and impairing food transportation. Browning, first at the edges and then of the entire leaf, is the result.
3. Some species deposit their eggs on green twigs, causing extensive damage (Color Figure 54).
4. Many species carry disease organisms that endanger plants.
5. Some species stunt the growth of plants by inhibiting growth on the under surfaces of the leaves, the places where they feed.

Proper insecticides can control leafhoppers. Spray foliage thoroughly and repeat if necessary.

Mites

Mites are very tiny animals rather than insects. They are usually no larger than the head of a pin. Mites have enormous repro-

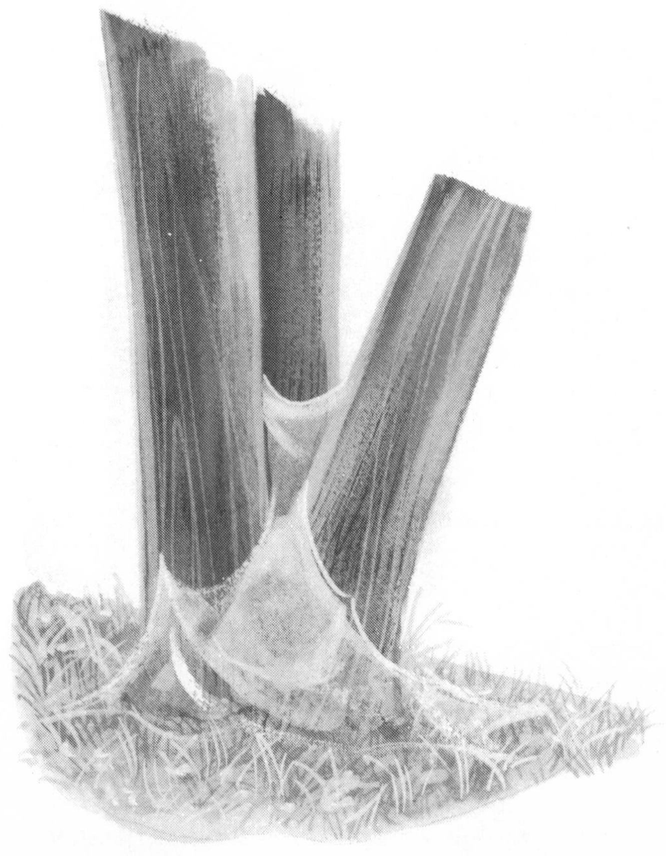

Mites Webbing

ductive powers and can remove large amounts of the essential plant juices when they attack in great numbers. They like a wide variety of plants, but are especially destructive to evergreens.

When the weather turns hot, plant leaves infested with mites will become blotched on both the upper and lower leaf surfaces (Color Figure 56). Eventually the leaves turn pale, die, and drop off. If you examine them closely, you will notice that they look as though they had been sprinkled with a fine, dirty white powder. Determining whether your plants are infested is easy. Hold a piece of white paper under a branch with one hand and slap or jar the branch with the other. The mites will drop down and you will be able to see them crawling around.

Plant Galls

Galls look like warts, swellings, and knots on the leaves, twigs, and branches of trees and shrubs. There are hundreds of kinds of galls, large and small, conspicuous and inconspicuous, but each kind is characteristic of the organism responsible for it. A gall-producer is choosy and remains faithful to its favorite plant.

Many of the common galls are due to abnormal cell growth stimulated by insects and mites, but some galls are caused by bacteria, fungi, and cylindrical worms. Still others are the result of the activities of aphids and fly larvae. The growth of many galls, particularly on oak trees, is stimulated by a number of species of small wasps.

One of the most frequent galls is the maple bladder-gall, which is usually found on silver maple leaves. It is caused by mites. Galls on hackberry leaves are caused by young plant lice, while midges (tiny mosquito look-alikes) are responsible for vein pocket gall on oak leaves and pod gall on honey locust leaves. The gouty oak gall, the wool-sower gall, and the oak apple gall are also common.

The habits of the gall-producer are as varied as the galls. In general, galls provide homes for insects, places to feed, lay eggs, and develop. In the case of oak galls, small wasps lay eggs on the buds and shoots. These hatch into legless grubs that cause the galls to develop. Afterwards, the galls are deserted by the insects.

Since galls are ugly, but rarely harm a tree (indeed, they sometimes provide protection against damage-causing insects), chemical control is really unnecessary. Control is justified, however, in the case of a young tree being stunted by the leaf damage caused by galls. By the time galls are easily observed on the leaves, it is too late to achieve any measure of control. Sprays must be applied when the insects or mites are still in the crawling state and are unprotected.

Scales

If you have a few spare minutes in the early spring, take a walk around your property and check your trees and shrubs. Many will be bare of leaves, making it

Oystershell Scale

easier to spot small abnormal growths resembling seashells tightly attached to the twigs or branches. These growths may be destructive scale insects (Color Figure 57). These insects are small, and all too often they are overlooked until after they have killed an entire branch. Even then, the scale is hard to recognize because it is cleverly camouflaged to look like part of the bark.

Scales come in a wide variety of shapes and forms. The growth is a protective shell under which the insects feed, lay eggs, and die. They infest twigs, branches, trunks, and fruit, depending on the species of scale and the host.

Scale populations are quick to increase in size. Growth retardation or death of the plant can occur before you know it. Thousands of scales sucking the sap from a plant cause the damage.

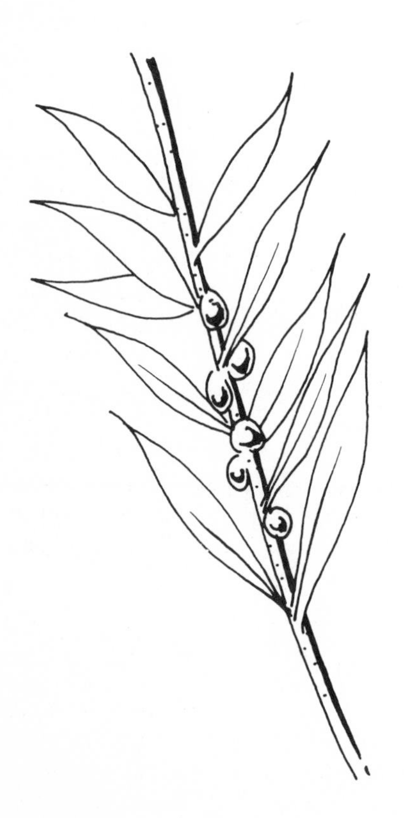

Fletcher Scale

Good control can be achieved during the period when the plant is dormant. Use an oil diluted according to label directions to smother the insects. These oils, however, should not be used on evergreens.

Insecticides used at the time when the scale insects emerge from their shells will give almost 100% control. A compressed air sprayer provides good coverage for shrubs.

Tree Borers

The many kinds of borers that attack trees and shrubs are the larvae of moths and beetles. Holes in the bark that ooze sap, sawdust, and insect debris and excrement indicate the presence of borers. They are particularly destructive to newly planted trees whose root systems have not adjusted. A number of contributory factors weaken trees and invite borers:

Drought—This is probably the primary factor that makes trees susceptible to borer attack. The dry or injured root system caused by drought systematically weakens the tree. A few seasons of low rainfall are usually followed by borers.

Construction—Changes in drainage and damage caused by digging and heavy machinery weaken trees considerably.

Extreme Weather—Hurricanes, ice storms, and extensive frost also weaken trees.

Sun Scald—A tree accustomed to a shady environment can be damaged by sudden exposure to the sun. This situation often occurs when a nearby tree is cut down.

Defoliation—Leaf-eating insects can damage a tree severely.

Chemical Injury—Salt used to melt snow, gasoline, oil, fumes from industry, and excessive amounts of fertilizer all cause serious injury.

Although many borers attack various sorts of plants, some are specialized, preferring only one type of tree or shrub. These picky eaters are easily identified by their choice of food. Among the most common trees that attract these very particular insects are the ash, azalea, box elder, bronze birch, cedar, cherry, chestnut, cottonwood, dogwood, hemlock, iris, lilac, linden, and peach.

The Plants

In the section that follows, the most common ornamental plants are listed alphabetically. Under each host plant heading, you will find the pest problems most often associated with that particular plant. Because aphids, galls, Gypsy moths, Japanese beetles, leafhoppers, mites, scales, and tree borers attack a wide variety of plants, their basic life cycles have already been discussed in more detail. However, the more important of these creatures will be included with their host plants.

There are so many thousands of species, some of them invisible to the naked eye, that you will probably not be able to recognize the pest by its physical characteristics. For this reason, the descriptions concentrate on the kinds of damage it causes. This is your surest guide to identification. Learn what varieties of trees and shrubs you have, and then read the list under those plants to find the pest most likely to have caused the injury you've noticed.

Control methods for handling the pests are listed in the chart at the end of the section. The life cycles of some pests vary from locality to locality, making generalized spraying instructions impossible. In these instances, we have recommended checking with your local Cooperative Extension Service, which will give you advice appropriate to your community.

Ailanthus (Tree of Heaven)

Ailanthus Webworm

This olive-brown caterpillar constructs a thin web and then feeds on the leaves of the ailanthus. It also eats the leaf stalks, causing them to wilt. Control measures are probably not needed unless the larvae are very numerous.

Cynthia Moth

This striking caterpillar is a leaf-eater. It is easily identified by its large size (about three inches in length) and dramatic coloring (green with black dots and blue knobs on the back). The adult is a large brown moth with a wingspan of about seven inches. There is no known method of control.

Arborvitae

Arborvitae Leafminer

The caterpillar of this species chews on the inside of the tree's needles, changing their color to white, then to tan, and finally to brown. The caterpillar is green, tinged with red.

Bagworm

During the winter and early spring, you may see spindle-shaped bags, one to two inches long, hanging from your trees and shrubs. These bags contain from 500 to 1000 bagworm eggs. When the newly hatched worm leaves the mother bag, it begins to feed on nearby foliage. It then constructs its own bag from silk threads and bits of foliage from the host plant. The appearance of the bag varies, therefore, from plant to plant. The adult males are black and the wingless adult females are yellow (Color Figure 44).

This defoliating insect usually does not kill deciduous trees (those that lose their leaves in the fall), but evergreens may succumb. You must spray while the worms are still small. The larger the worms, the harder they are to control. Once the worm stops feeding, spraying becomes ineffective.

Juniper Scale (see, Juniper)

Spider Mites (see, Spruce)

Ash

Ash Borer (see also, Tree Borers)

This insect deposits eggs in the cracks and crevices in the tree bark or scar tissue or just below the soil line. The tiny hatching larvae quickly drill through the bark and enter the tree. Sprays should be applied when the adults are in the process of laying eggs.

Fall Webworm

The fall webworm, alias the fall cankerworm, the fall measuring worm, and fall inchworm, covers leaf surfaces with a layer of silk, eventually binding the branches together. Because the trees become dormant in the fall when they appear, no control is usually necessary.

Oystershell Scale (see, Scales)

Red-Headed Ash Borer (see also, Tree Borers)

Timber! These little grubs can girdle a tree so badly that it will topple. The adults are reddish-brown beetles that attack hardwood trees such as ash, hickory, and oak. No way of controlling their temper tantrums is known.

Birch

Birch Leafminer

The larva mines its way between the upper and lower surfaces of the leaf, making it look blistered. Check the leaves early in the spring because control measures must be taken when the mines first appear.

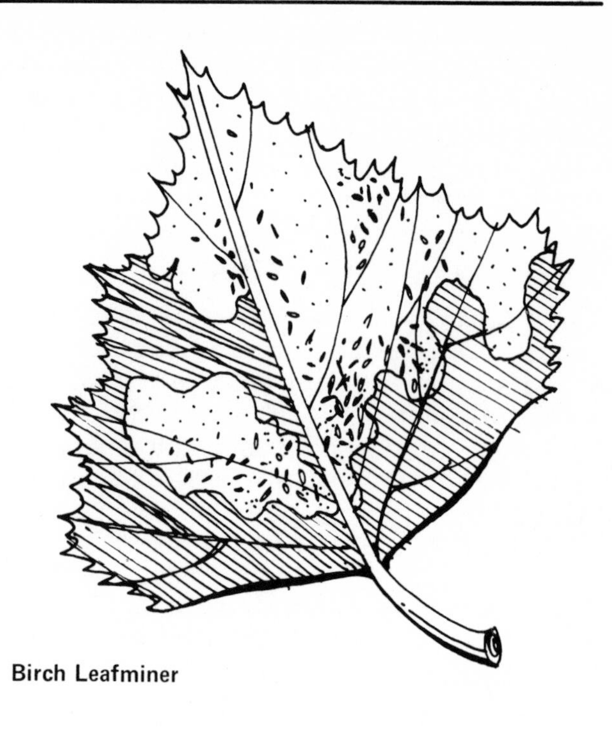

Birch Leafminer

Birch Skeletonizer

These tiny caterpillars chew their way into the leaf and feed for a few weeks. They then skeletonize the bottom part of the leaf and drop to the ground where they spend the winter in cocoons. Most of the damage occurs in August. Control spraying should be done when the insects are first noticed.

Bronze Birch Borer (see also, Tree Borers)

This insect is deadly to birch trees, killing them in two or three seasons. The eggs are laid in the rough crevices of the bark on the upper part of the tree. Some swelling may be noticeable. The adult beetles emerge from the upper limbs of the tree. Trees infested with this insect have a dead area at the top which will continue to spread unless control measures are taken (Color Figure 55).

Thorough spraying as the eggs begin to hatch is helpful. The first spraying should be done in late spring and the application repeated two weeks later. A promising new line of research using a small parasitic wasp to control the borer naturally is now being explored.

Box Elder

Boxelder Aphid (see also, Aphids)

Once in a while, these green, hairy aphids experience a population explosion. When this occurs, there may be severe injury to the tree. Their sticky secretion, called "honeydew," covers sidewalks, parked cars, and anything else under the tree. Spray only if damage is extensive or if the number of aphids makes them a nuisance.

Boxelder Bug (see, "People Pests")

Eastern Tent Caterpillar

These black, hairy caterpillars construct tents on the forks or tips of branches in the spring (Color Figure 47). They feed during the day and retreat to their tents at night. This stage continues for five or six weeks. The adult moths, which are reddish-brown, emerge in the early spring. Every ten years or so, their numbers increase dramatically.

Sprays should be applied when the tents first appear, and areas infested with tents should be pruned. Do not attempt to burn the insects out with gasoline-soaked rags attached to the end of a pole. This folk remedy can be fatal.

Tent Caterpillar

Boxwood

Boxwood Leafminer

The first signs of these pests are small blotches on the leaves. These enlarge into blisters that cause the leaves to turn brown and then drop off. If you look carefully, you may be able to observe small yellow-green worms feeding on the leaf tissue. The adults are orange-yellow midge flies.

Catalpa

Catalpa Sphinx

This exotically named insect sounds like it should be found in an Egyptian pyramid. In reality, however, it is a voracious caterpillar that feeds on catalpa foliage. One larva can consume an entire leaf. It is pale yellow with green markings. Treatment is necessary only when the insects are numerous.

Comstock Mealybug

You are most likely to find this white, waxy creature in cottony masses at the forks of tender young shoots or near the base of a leaf. It is a sucking insect like the aphid and may produce a few generations each year. Unless the insects are numerous, control measures are not needed.

Crabs and Hawthorns

European Red Mite (see also, Mites)

If the leaves of your crab or hawthorn trees have lost their glossy green color and appear brownish, dry, and curled, and if you can't find any visible cause, you should suspect mites. This tiny species is red and about the size of a pencil dot. You may find molted skins in the leaf curls. The young mites rasp the leaf surface and remove the chlorophyll and plant juices.

Hawthorn Lace Bug

A small, dark-colored insect, the hawthorn lace bug attacks a number of trees besides the hawthorn. Among its favorites are the juneberry, the buttonbush, and the quince. This bug sucks out plant juices, causing the leaves to become brown and spotted.

Hawthorn Leafminers

Hawthorn Leafminer

The larvae of this species mine between the sheaths of the leaves, creating brown dead areas at the tip of the leaf (Color Figure 52). If you hold the leaf up to a light, you'll see black fecal pellets left by the larvae. The adults are bee-like insects called sawflies. Sprays should be applied in the spring.

Hawthorn Mealybug

This is one of the most unforgettable insects you'll ever encounter. It is oval-shaped, segmented, soft-bodied, and covered with a white powdery or cottony substance. Mealybugs are sucking insects that feed on plant juices. They can be a real problem on houseplants, but seldom require controls on outdoor plants.

San Jose Scale (see, Scales)

Wooly Hawthorn Aphid (see also, Aphids)

Like many other aphid species, the wooly hawthorn aphid sucks juices out of the leaves, making them shrivel. These white plant lice move about in great numbers on the twigs and branches and are easily seen. Unless the insects are numerous, spraying is unnecessary.

Yellow-Necked Caterpillar

These fuzzy reddish caterpillars are colorful creatures with yellow necks, black heads, and bold yellow racing stripes along their sides. They love the foliage of fruit trees.

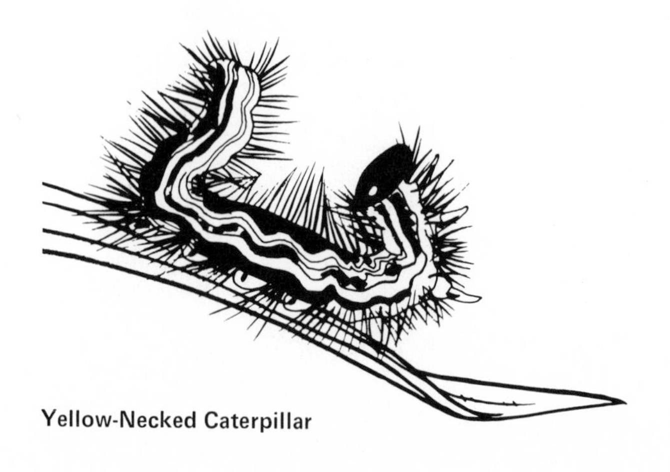

Yellow-Necked Caterpillar

You'll find them feeding in bunches. When disturbed, they rear up en masse, rather like the Radio City Rockettes. Control spraying is not necessary unless the insect population is very large.

Dogwood

Dogwood Borer (see also, Tree Borers)

The end branches of the dogwood, one of our most beautiful shade trees, are often infested with this pest. The borer enters the tree by burrowing through the bark, especially in areas where the tree has been damaged in the past. This insect can cause the loss of several branches or the death of the whole tree. Insecticides must be applied in late spring or early summer. Several applications are necessary.

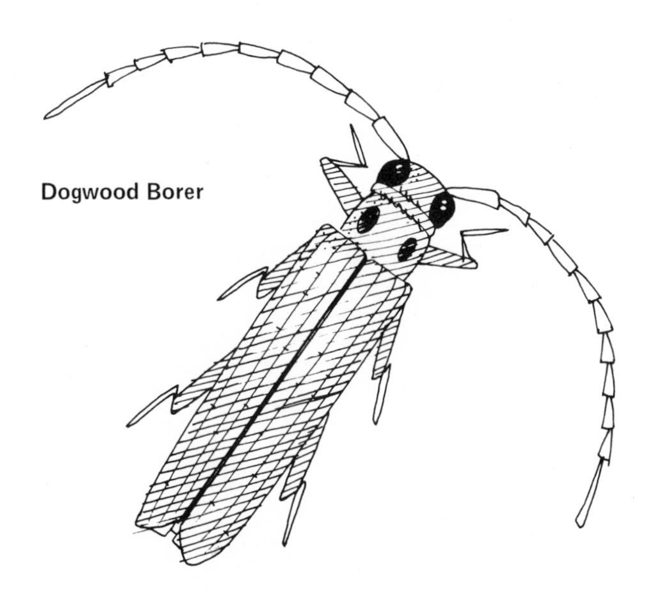

Dogwood Borer

Dogwood Scale (see, Scales)

Douglas Fir (see, Plant Galls)

Elm

Banded Elm Leafhopper (see also, Leafhoppers)

Phloem necrosis, a virus disease of epidemic proportions among American elms,

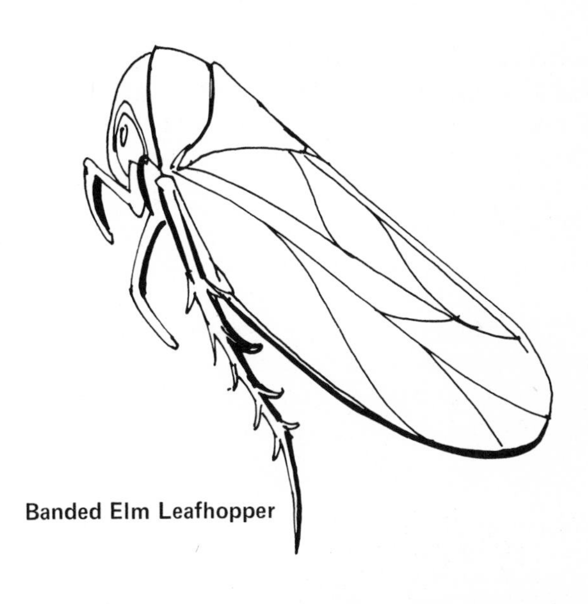

Banded Elm Leafhopper

is carried by this insect. The immature leafhopper feeds on young foliage, creating brown areas where it sucks the sap. Spraying is usually not necessary.

Elm Borer (see, Tree Borers)

Elm Cockscomb Gall Aphid (see also, Aphids and Plant Galls)

This gall is located on the elm leaves and

Elm Cockscomb Gall Aphid

gets its name from its resemblance to a cock's comb. The gall is produced by a species of aphid that sets up housekeeping on the leaf. No control is needed.

Elm Leaf Aphid (see also, Aphids)

An offensive little insect, the elm leaf aphid is a real "pain in the neck" because it secretes "honeydew," a sticky substance that drops down onto anything left under the tree. Although it can be a nuisance, damage to the tree is very rare.

Elm Leaf Beetle

Both the larvae and adults of this insect are leaf-eaters. They prefer Chinese elm, but will settle for any species. If your elm leaves are skeletonized or dried out, or if they curl and drop prematurely, this insect is probably the culprit. The larva is dark, almost black, and the adult is a beetle about one-fourth inch long with yellow legs and a yellow-green body. Treat when the insects are numerous.

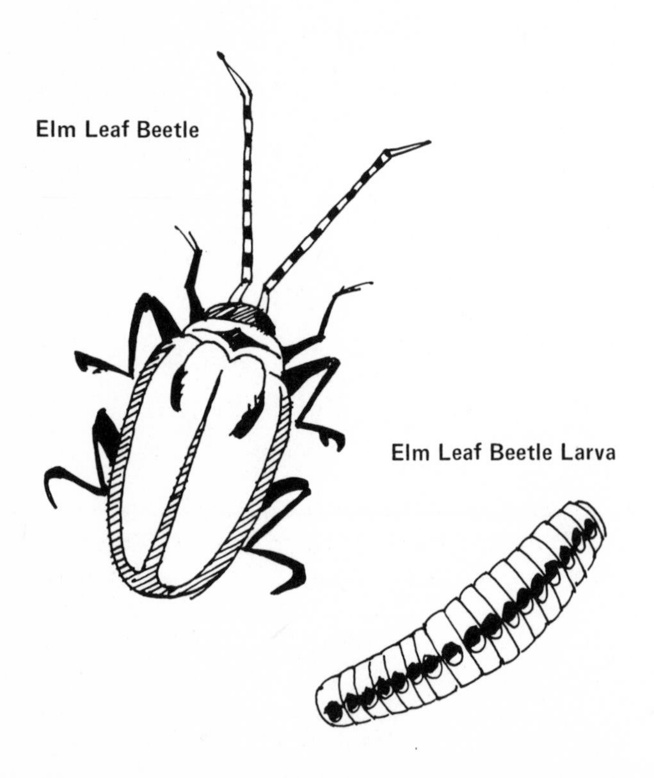

European Elm Scale (see also, Scales)

If your elm has yellowing leaves or leaves that are dropping prematurely, this insect has probably attacked. Twigs die and occasionally an entire tree may succumb. The wingless females do most of the damage. When very young, they move freely about the tree. Later they insert their beaks into the tree, suck out the juices, and remain in that spot, looking like small bumps, until they die. Sprays should be applied in the eary spring when the leaves are beginning to form.

Fall Webworm (see, Ash)

Mites (see also, Mites)

A few species of mites can be found on elms, but they inflict little damage. Once in a blue moon, a heavy mite population may turn the leaves brown.

Putnam Scale (see, Scales)

Scurfy Scale (see, Scales)

Smaller European Elm Bark Beetle

Don't let this name fool you! This devastating little insect is responsible for Dutch elm

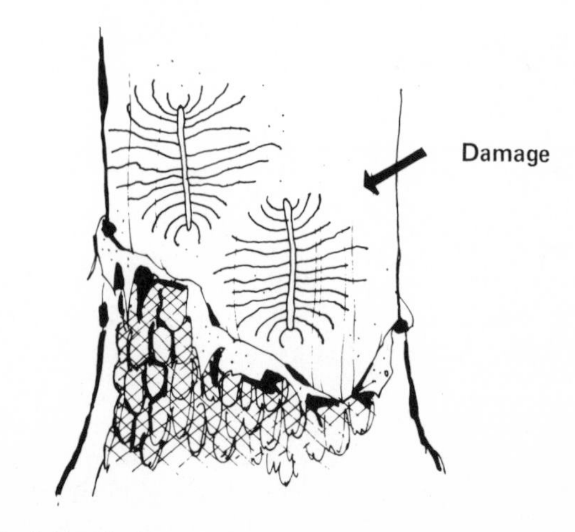

Elm Bark Beetle Damage

disease. Look closely at the bark of your elm tree. If you see hundreds or even thousands of little round holes (about the size of a BB pellet), this pest has moved in. The reddish-brown beetle carries the disease fungus from infected to healthy trees.

Spring Cankerworm (see, Cankerworms)

White-Marked Tussock Moth

This striking caterpillar is covered with tufts of long black and white hairs that serve as a protective mechanism. If you try to pick one of these insects up, the hairs, which contain an irritating chemical, will prick your fingers and cause a burning sensation. The larvae are ravenous feeders who skeletonize foliage. Spraying is helpful when the population is large.

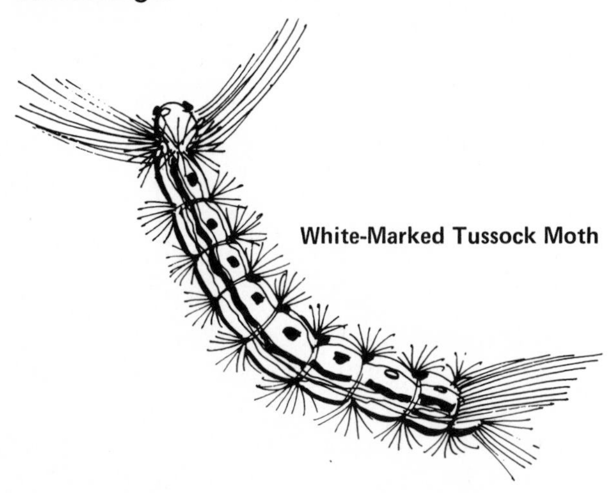

White-Marked Tussock Moth

Wooly Elm Aphid (see also, Aphids)

Curled leaves may indicate the presence of these dark-green to black plant lice. They love to suck juices out of the underside of the leaf. Treatment is indicated only when the larvae are numerous.

Euonymous

Euonymous Scale (see also, Scales)

This is "the" pest of the euonymous. Look for tiny white specks on the leaves. This insect spreads quickly, but control is usually not necessary.

Winged Euonymous Scale (see, Scales)

Hackberry

Hackberry Nipple Gall Psyllid (see also, Plant Galls)

Nipple gall is easily identified and its name makes it easy to remember. The galls resemble nipples and grow on the underside of the leaf.

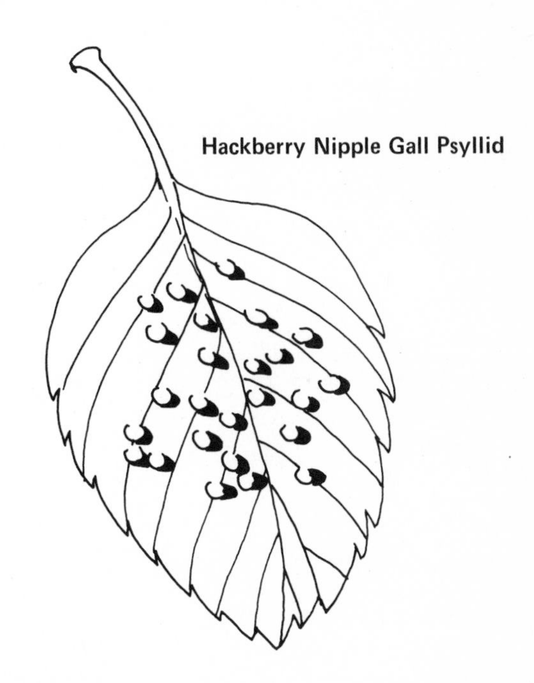

Hackberry Nipple Gall Psyllid

Witches'-Broom

You may notice branches growing in abnormal bunches from the main branch. The formation resembles a broom. This growth may be caused by mites, a fungus, or both—no one is really sure. In any case, it looks eerie, but will not harm your tree.

Hickory

Hickory Bark Beetle

Check for small beetle exit holes in the bark. If the tree is heavily infested, it will probably die and should be cut down and burned or buried during the winter months in order to destroy the larvae (grubs) in the tree. Keep your healthy trees well watered and fertilized.

Hickory Gall Phylloxera or Aphid (see also, Aphids and Plant Galls)

A heavy population of this creature can cause problems. Keep your eyes open, since this usually isn't noticed until too late. The galls are produced by aphids that live on new twigs and prevent normal growth. The twigs eventually die from the gall outward. Treatment must be administered when the leaves first begin to break.

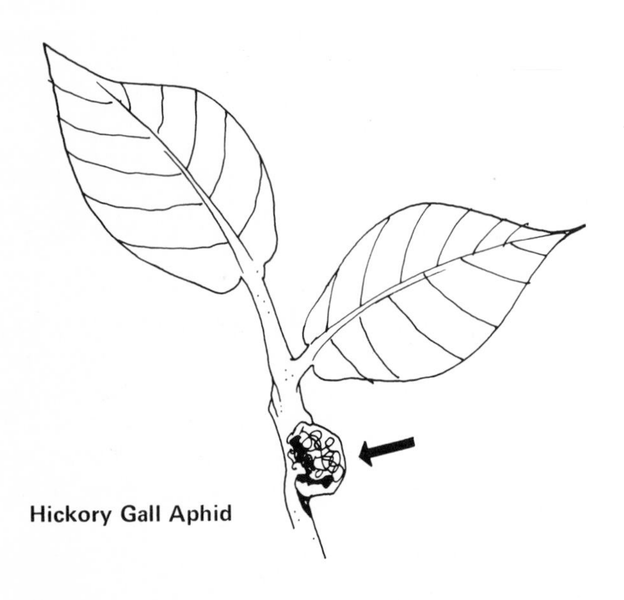
Hickory Gall Aphid

Hickory Horned Devil (Regal Moth)

This sounds like a great title for a William Friedkin movie, but it is really a hungry insect that can provide a science project for a non-squeamish youngster. Capture one of the larva and place it in a Mason jar with some foliage from the tree. The larva is about five inches long and has a green body with a red head and curving red horns. After a few days, it will begin to spin a cocoon. Eventually (it may take weeks to months), a beautiful olive-green moth with a wingspan of four to six inches will emerge. Although the larvae love to eat, they are too few to cause much damage.

Yellow-Necked Caterpillar (see, Crabs and Hawthorns)

Holly

Holly Leafminer

This insect, like the other mining species, blisters the leaves. If the infestation is heavy enough, the shrub will look disfigured and the leaves will drop prematurely. Sprays should be applied when the mines first appear.

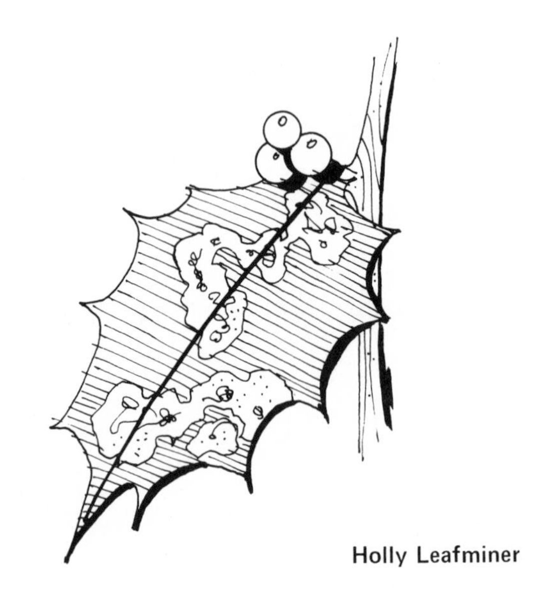
Holly Leafminer

Juniper

Bagworm (see, Arborvitae)

Juniper Bark Beetle

This insect doesn't really bark—in fact, it probably never makes a sound. What it does do is eat. It feeds in the crotches of small branches, weakening them so that they bend downward until they break and die.

Juniper Scale (see, Scales)

Juniper Webworm

These light-brown caterpillars build webs that bind the juniper needles together. So, look for webs and brown foliage. The adult moths lay their eggs in the leaf axils of new growth.

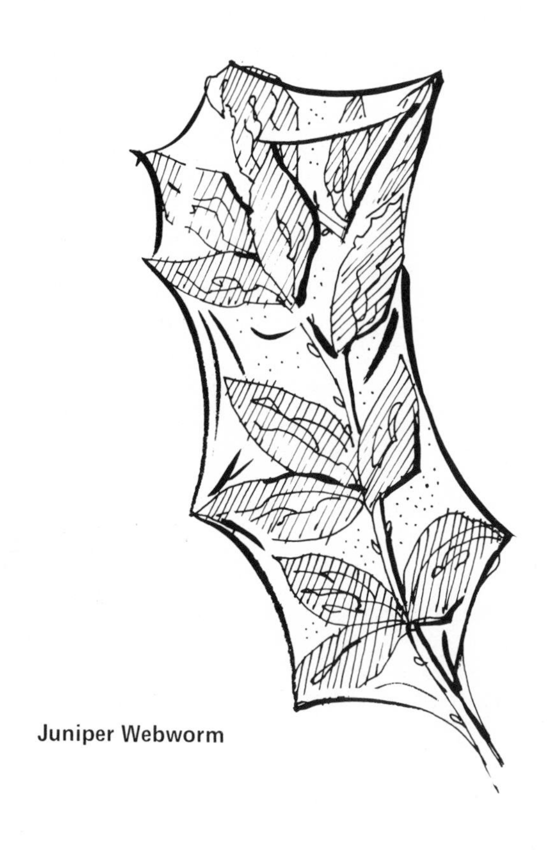

Juniper Webworm

Spruce Spider Mite (see, Spruce)

Lilac

Lilac Borer (see also, Tree Borers)

This is the major lilac pest. The caterpillar is creamy white and is about three-fourths inch long. The adult lays its eggs on

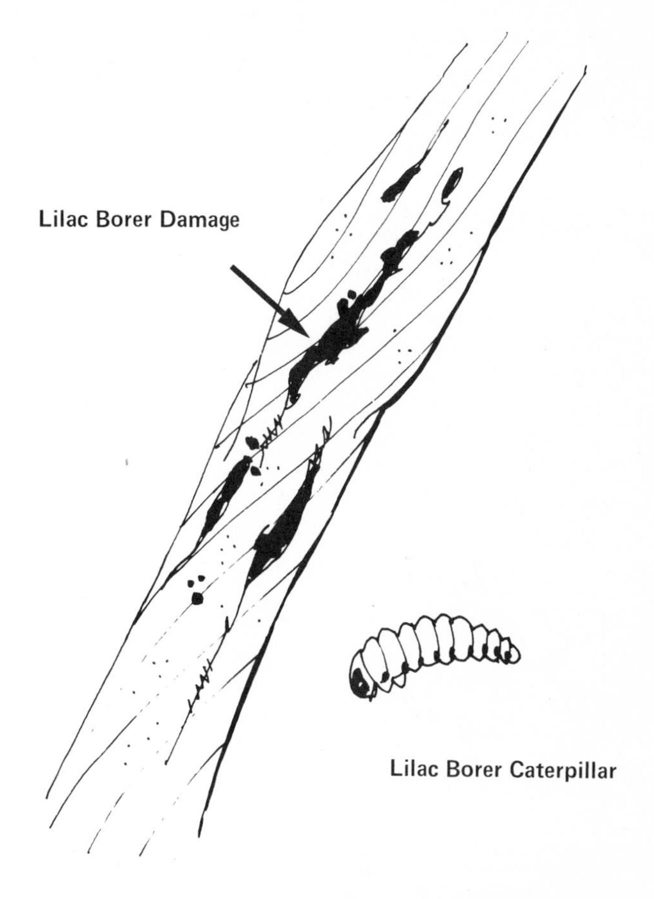

Lilac Borer Damage

Lilac Borer Caterpillar

roughened or wounded areas of the shrub. After hatching, the larvae tunnel under the bark and burrow into the wood, weakening the stem and causing the foliage to wilt. Borer holes can be easily seen on the older stems of the shrub where borers have worked for several seasons.

Oystershell Scale (see, Scales)

Linden

Elm Spanworm (Snow-White Linden Moth)

These insects cause great devastation every 10 or 20 years, but you'll probably only encounter them once in your lifetime. They appear in such vast numbers on these occasions that the leaves are stripped not only from the linden but from other species of trees as well. The worm is about 1½ inches long and is dark brown marked with

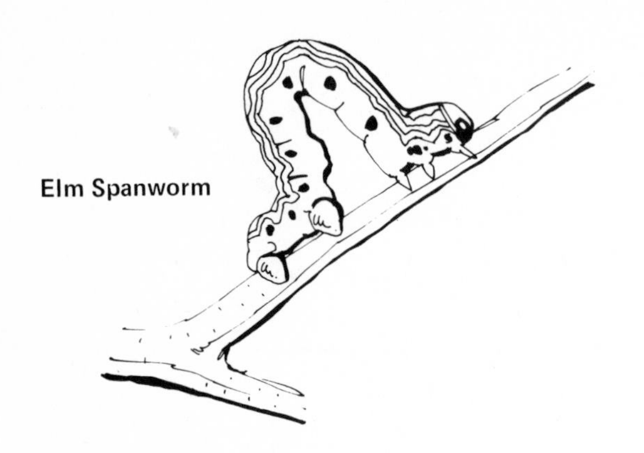
Elm Spanworm

red. Control measures are usually unnecessary except during periods of heavy infestation.

Locust

Bagworm (see, Arborvitae)

Honey Locust Pod Gall (see also, Plant Galls)

The adult midges usually begin appearing in April, about the same time the locust tree starts its growth. They're only one-eighth inch long. The female lays tiny, kidney-shaped eggs that vary in color from lemon-yellow to light amber. The larvae hatch out in a day or two and immediately start feeding. They eat the inner surface of the leaflet, stopping its growth, but the outer surface continues to grow normally. This produces a pod. When the larvae finish feeding, they change to the pupal (cocoon) stage in the pod. The emerging adults escape between the leaflets, leaving their pupal cases behind. If conditions are right, up to seven broods can occur in a single season. Despite this activity, no control is necessary.

Honey Locust Pod Gall

Locust Mite (see also, Mites)

Mites are so tiny that you'll probably never see one. Look for signs of their presence instead—stripping, blanched leaves, and/or premature leaf drop. Control sprays must be applied in early spring when the leaves are small.

Mimosa Webworm

The mimosa webworm has been an important pest in the Washington, DC, area since the early 1940's. In recent years, it has spread to the Midwest. Look for webbing, drying, skeletonizing, or browning of the leaves. Sprays should be applied when the webs first appear.

Magnolia

Magnolia Scale (see, Scales)

Maple

Aphids (see also, Aphids)

These pests have been mentioned a number of times. Although there are hundreds of species of aphids, two—the Norway maple aphid and the painted maple aphid—are particularly fond of maple trees. Both are sap-suckers that secrete "honeydew." Controls are only needed when the population is heavy.

Cotton Maple Leaf Scale (see also, Scales)

These white, cottony, fibrous egg masses are found on leaves. Look for premature leaf drop and dead twigs.

Cottony Maple Scale (see also, Scales)

These are similar in appearance to the cotton maple leaf scales, but the white popcorn-like balls are strung along the branches (Color Figure 48).

Eriophyid Mite (see also, Mites)

These plant-sucking creatures love the tender young leaves of trees, especially maples. They cause brilliant purple, pink, or red blister-like growths or blotches. Control measures are generally unnecessary, but injury to the tree may require the application of a miticide. Ordinary insecticides should not be used because mites are not insects.

Flat-Headed Apple Tree Borer (see also, Tree Borers)

It may seem odd to list a creature with this name under maples, but this insect is fond of all kinds of shade and fruit trees, and maple is one of its favorites.

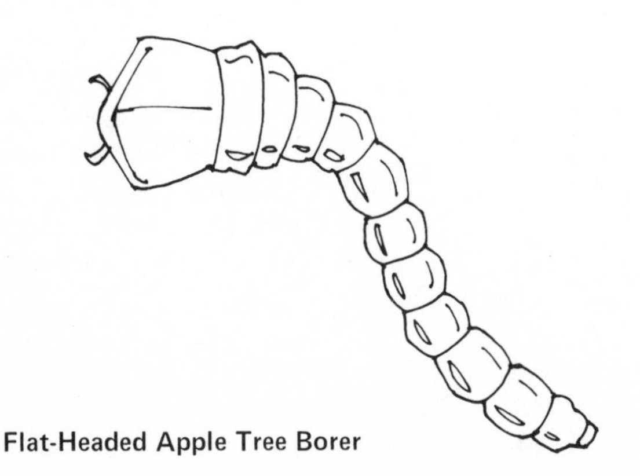

Flat-Headed Apple Tree Borer

Green-Striped Mapleworm

This is another leaf-feeder. The yellow-green caterpillars have red heads and are about 1½ inches long. They are ravenous feeders, often defoliating trees twice in one season. Controls are not usually needed.

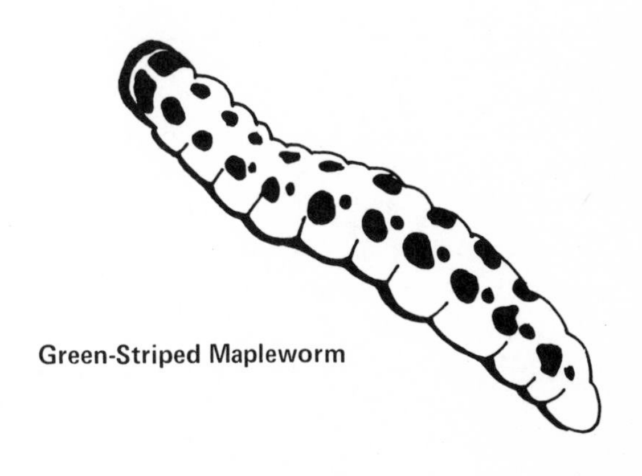

Green-Striped Mapleworm

Maple Bladder-Gall Mite (see also, Mites and Plant Galls)

Look for little green, red, or black growths on the top side of the leaf. These ugly growths indicate maple bladder-gall (Color Figure 49). Even though they are unattractive, they don't require control measures.

Maple Bladder-Gall

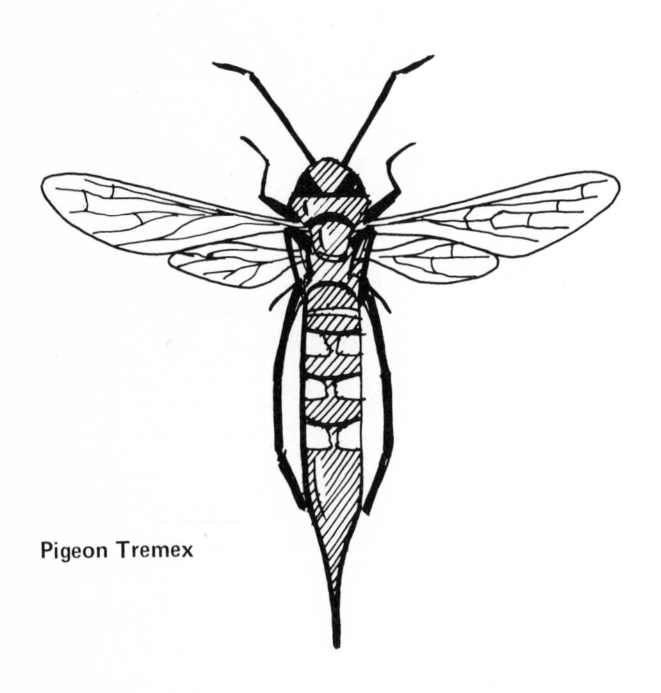
Pigeon Tremex

Pigeon Tremex

This oddly named creature is not a bird, but an insect. This two-inch-long sawfly bores holes into diseased maples and other trees. It is sometimes blamed for the damage, but it is really only taking advantage of preexisting disease. Artificial controls are not needed. There is a slender parasitic wasp that drills into the pigeon tremex tunnels to deposit its eggs. The wasp larvae attach themselves to the pigeon tremex larvae (grubs) and kill the little devils, taking care of your problem for you. How about that!

Oak

Although a large number of insects can create problems for the oak, these trees seldom succumb to disease or insects. Most deaths can be attributed to urbanization and the resulting destruction of the delicate ecological balance the oak needs to survive.

Carpenterworm (see, Poplar)

Forest Tent Caterpillar

This insect spins silken threads, but doesn't make a real tent. Although it prefers oak and maple trees, it manages to live happily on apple, birch, hawthorn, peach, pear, plum, and willow trees as well. The larvae, which are 1½ inches long, are black marked with white spots and yellow stripes. They gather in large groups on tree branches and eat the young leaves. The adults are brownish-buff moths that emerge in late June or early July. Insecticides can be very effective when the caterpillars are young.

May Beetles

Sometimes called June bugs, these large brown beetles think oak leaves are yummy (Color Figure 60). Small trees can be entirely defoliated in a short time. The larvae feed on grass or vegetable roots. Insecticides should be used when the infestation is extensive.

Oak Borer (see, Tree Borers)

Oak Gall Insects (see also, Plant Galls)

Oak apple gall (Color Figure 46), gouty oak gall, and wool-sower gall rarely do serious harm, but their names are enough to scare you!

Oak Kermes

This is a real lover of oaks—too bad the feeling isn't mutual. Bur oak is its favorite, but it also likes chinquapin, white, and red oaks. New shoots are smothered by clusters of kermes and end up distorted or dead.

Oak Leafminer

This group (Color Figure 42) includes a

variety of insects. The larvae of moths, flies, beetles, and sawflies diligently tunnel between the upper and lower leaf surfaces. Their finished work resembles a blister.

Periodical Cicada

The appearances of enormous numbers of cicadas (Color Figure 43) are often confused with the locust (grasshopper) swarms that occasionally devastate large areas and which were one of the plagues the Egyptians suffered in Biblical times. In 1889, a swarm of locusts covering 2000 square miles was observed. An estimated 250 billion locusts, weighing about 500,000 tons, devoured everything in sight.

The periodical cicada, often called the 17-year locust, is not really a locust at all. It doesn't even eat. Whatever damage it does occurs during the egg-laying process when the female slits the tops of branches and twigs and deposits eggs in the tree tissue. The ends of these small branches turn brown and die. Severe damage sometimes occurs in newly planted trees. Although they don't injure evergreens, vegetables, or flowers, any new planting of shade trees or shrubs may attract them. They prefer oak, hickory, apple, peach, and pear trees and grape vines.

The males have high-pitched voices that can be heard for one fourth of a mile or more. This shrill sound, combined with the large numbers of cicadas, creates some nervousness in people close by. The cicada, however, does not bite or injure people or animals and poses no threat.

The periodical cicadas are orange to black in color and have large transparent wings. Their bulging eyes make them resemble frogs. You will see them in late May or early June. The annual cicada, which does little damage, is larger, green to black in color, and appears later in the season (July to September).

Cicadas can't really be controlled, but protective spraying may be helpful. Treat only if the cicadas are actually seen in your trees.

Pit-Making Oak Scale (see also, Scales)

A small pit develops where the scale settles. The depression is still noticeable after the insect's death. The twigs and branches become rough to the touch and occasionally die as a result of the extraction of the plant juices.

Twig Pruner

This insect is something of a gardener. It prunes twigs by cutting around them just under the bark. Look for almost perfectly cut twigs littering your yard during late summer. Shade trees such as oak, elm, hickory, and maple may lose their shapes from overpruning by this brown beetle. No method of control is known.

Pine

European Pine Shoot Moth

These insects attack young growth. The boring causes shoots to turn brown and die. Mugho, red, and Scotch pines are particularly likely to be bothered. This problem is easy to identify but hard to control.

Nantucket Pine Moth

Several varieties of pine, including red and jack pines, are attacked by this moth. The larvae knock off a few needles and then bore into the centers of the twigs. Look for dead tips and/or delicate webs surrounding the needles and twigs. Damage may be quite severe. The larvae are yellow to light brown in color and the adult moths are reddish-brown marked with silver

Pine Bark Aphid (see also, Aphids)

This creature loves white, Scotch, and Austrian pines. Look for white cottony masses on the trunks and branches. Older trees can usually laugh off this pest, but it may cause problems for young trees. Spraying can help protect them.

Pine Needle Scale (see also, Scales)

This scale turns mugho and Austrian pine needles brown. Check for white specks stuck to the needles (Color Figure 50). A large number of specks may indicate that spraying is needed.

Pine Tortoise Scale (see also, Scales)

Although Scotch pine is its very favorite, this unfussy scale will settle for jack, mugho, and Austrian pines as well. Heavily infested trees are coated with black mold and foliage and needles are severely affected. Young trees may die.

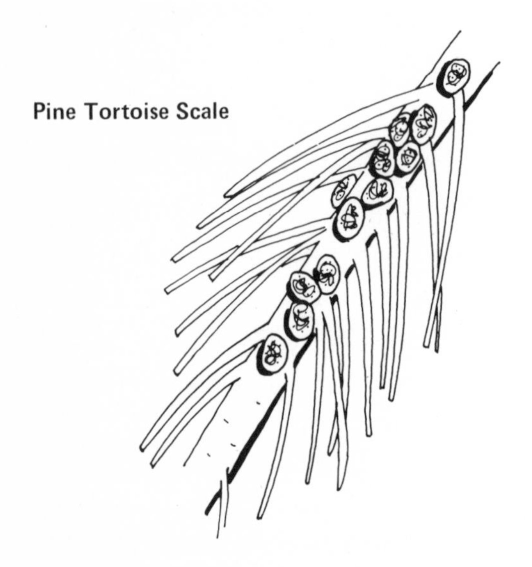

Pine Tortoise Scale

Red-Headed Sawfly

If you walk into your garden one day and find that your pine looks rather bare, you're probably seeing the results of the red-headed sawfly. These insects can defoliate a young tree, often killing it in the process. The larvae, which live in groups, are whitish with brown heads when young. As they mature, their color changes to yellow with black spots and their heads become red. Measures should be taken when the insects are numerous.

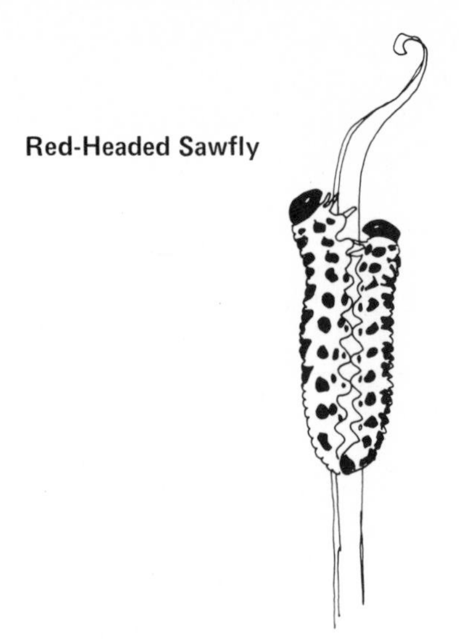

Red-Headed Sawfly

White Pine Weevil

This insect is a serious threat. Given a choice, it prefers white pines, but it can also be found on Scotch and jack pines. Look for dead and dying branch ends. There is no effective control method.

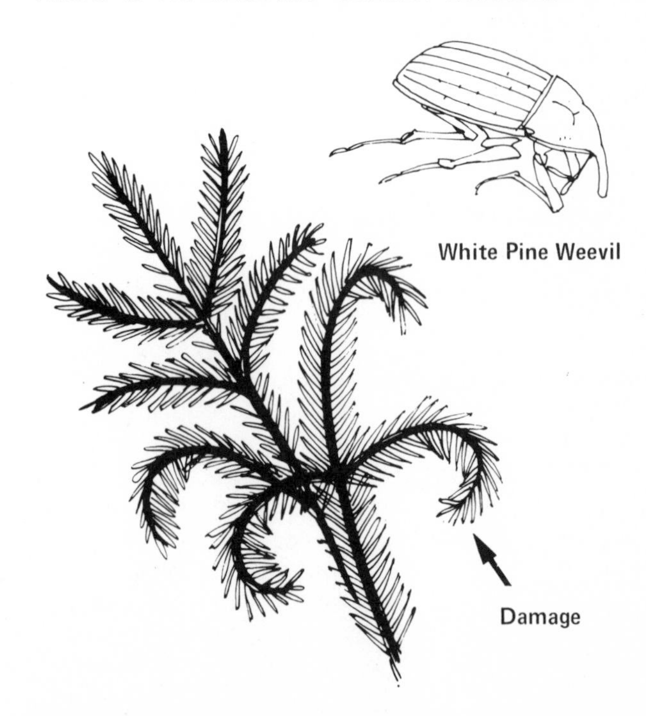

White Pine Weevil

Yellow-Bellied Sapsucker

This bird is mentioned here because it is often responsible for damage that homeowners blame on borers. It pecks holes in the bark and feeds on the sap that oozes out. Sometimes it pecks too many holes at once and some of the sap is left uneaten. When the sugar in the sap sets for a while, it begins to ferment. When the bird comes back and starts to feed, he actually gets drunk! Look for holes in a neat row.

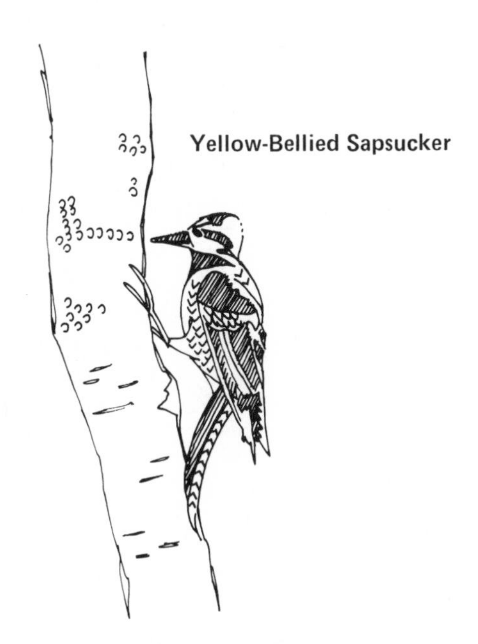

Yellow-Bellied Sapsucker

Zimmerman Pine Moth

If you look up into the tree, you will notice a gummy material where the branch and trunk join. This substance is the result of the larvae tunneling under the bark. Branches may turn brown and drop off.

Poplar

Carpenterworm

Despite its name, this insect is not a skilled craftsman. It gets its name from the large tunnels it bores into wood. Branches, and even whole trees, can be deformed by this process, which may take up to three years. The adult moth lays eggs in bark crevices, old tunnels, and wounds. There is no known method of control.

Cottonwood Borer (see, Tree Borers)

Cottonwood Leaf Beetle

This insect is a leaf skeletonizer, eating everything but the veins. Sprays should be used when the infestation is heavy.

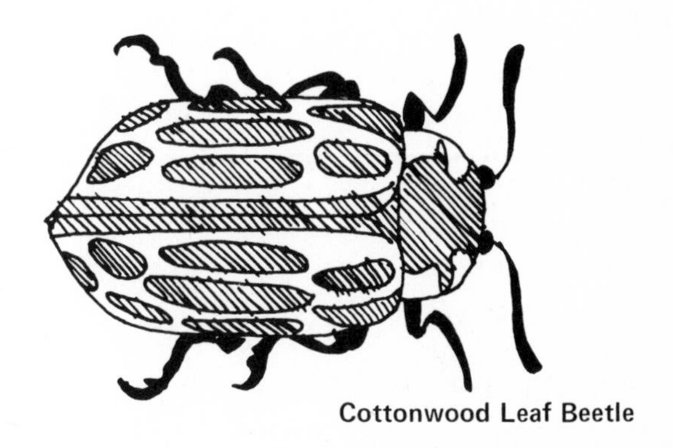

Cottonwood Leaf Beetle

Oystershell Scales (see also, Scales)

This insect (Color Figure 57) is seldom a pest in months ending with "r."

Poplar Borer (see, Tree Borers)

Poplar Tentmaker

This tentmaker should not be confused with the more famous Omar. These black caterpillars have yellow and brown stripes and are a little more than one inch long (Color Figure 47). They weave silken tents, enclosing branches and twigs. If you find tents on your poplar tree, these are the culprits. Unless they are numerous, treatment is not needed.

Privet

Privet Thrips

This insect likes privacy. It feeds on the underside of the leaves where it rasps and sucks out the chlorophyll. You won't see it until the leaves start to fall off. Stay alert, because this is one problem you won't find unless you look for it. If the infestation is heavy or if damage occurs, spraying can be helpful.

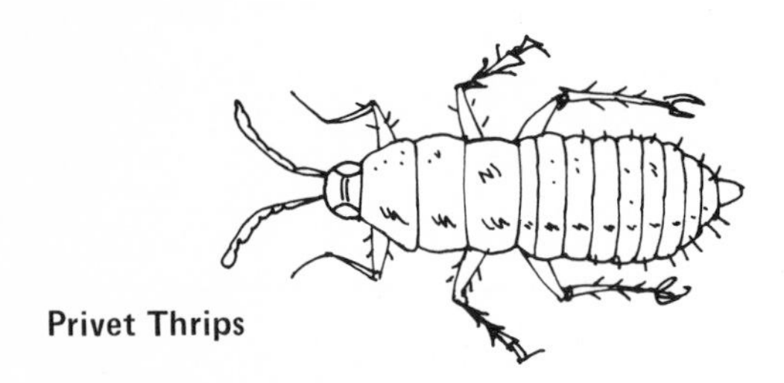

Privet Thrips

Redbud (Judas Tree)

Oystershell Scale (see, Scales)

Redbud Leafhopper (see also, Leafhoppers)

Leafhoppers are plant-suckers. They occur in such large numbers that when they are disturbed the tree appears to be surrounded by a cloud. Because they remove plant juices and chlorophyll, the leaves become blanched. A heavy infestation requires spraying.

Two-Spotted Spider Mite

These tiny creatures multiply faster than an electronic calculator. They consume so much chlorophyll and plant juice that the leaves shrivel and drop prematurely. If you find yourself raking leaves in mid-summer, you probably have spider mites. Use a miticide if damage is severe.

Spirea

Spirea Aphid (see also, Aphids)

These green plant lice produce great amounts of "honeydew." Look for them on tender new growth. Their feeding makes the leaves curl, eventually wrapping them around the insects. Spraying can be helpful if the population is heavy.

Spruce

Cooley Spruce Gall Aphid (see also, Aphids and Plant Galls)

This insect is not a Chinese import, despite is intriguing name. It is found on Colorado blue, Sitka, and Engelmann spruces. The gall looks like a pineapple and can be seen on the tips of the twigs.

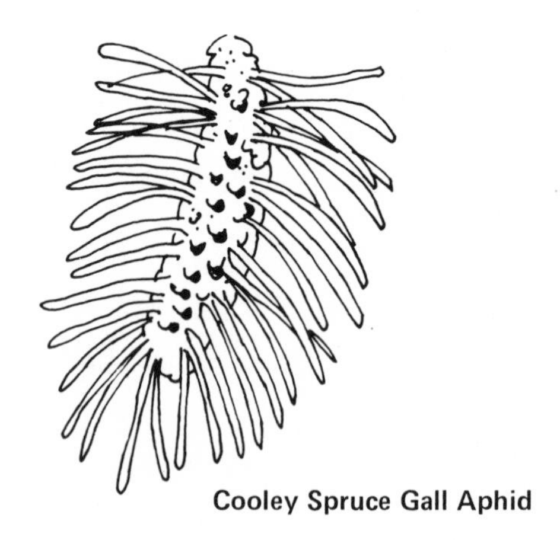

Cooley Spruce Gall Aphid

Eastern Spruce Gall Aphid (see also, Aphids and Plant Galls)

This insect is similar to the cooley spruce gall aphid, but its pineapple-shaped galls are found at the base of the twigs rather than at the tips.

Pine Needle Scale (see, Pine)

Spruce Bud Scale (see also, Scales)

These mahogany-colored, round insects cluster in small groups around the nodes and buds of spruce trees. They can be difficult to detect because they look so much like the buds. Spray when you see the young crawling around.

Spruce Budworm

These insects feed on needles and opening buds. You probably won't see many of them because they prefer forests to ornamental trees.

Spruce Spider Mite

These very small mites can be a serious threat to evergreens. Quick intervention is necessary once the tree starts turning brown or you may find yourself doing a Paul Bunyan imitation.

Sweetgum

Sweetgum Scale (see also, Scales)

Some leaf damage is caused by this creature, but it is seldom severe enough to worry about.

Sycamore

Bagworm (see, Arborvitae)

Sycamore Borer (see, Tree Borers)

Sycamore Lace Bug

This delicate little insect is a real lush. It removes tremendous amounts of plant juice and chlorophyll. The leaves become littered with its waste products, and eventually lose color and dry up. If your tree looks pale and dry, suspect the sycamore lace bug. Treatment is needed when the population is heavy.

Tulip Tree

Tulip Tree Aphid (see also, Aphids)

As this aphid marches across the tulip tree, he drips "honeydew." A sooty mold grows on this secretion, making leaves and branches turn black. Check for curled, drooping leaves with a dirty appearance.

Tulip Tree Scale (see also, Scales)

Usually found on the lower branches and twigs, this insect, like the other scales, is a plant-sucker.

Walnut

Black Walnut Curculio

Crescent-shaped cuts indicate the presence of this pest. The pale red curculio is especially fond of young, tender shoots. There is no known method of control.

Walnut Caterpillar

A real nut fancier, this caterpillar feeds on walnut, hickory, pecan, and butternut trees, but can sometimes be found on fruit trees as well. It is about two inches long. When newly hatched, it is reddish, but as it grows, its color changes to black with a covering of long white hair. This insect likes to begin feeding at the tip of a branch, gradually working its way down to the base.

Sycamore Lace Bug

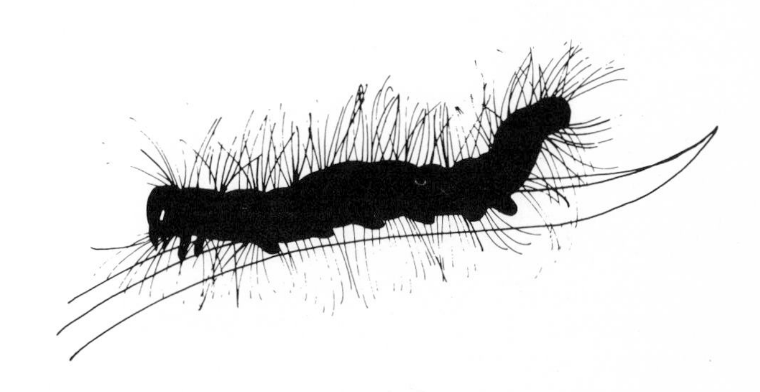

Walnut Caterpillar

Willow

Bagworm (see, Arborvitae)

Oystershell Scale (see, Scales)

Willow Aphid (see also, Aphids)

There are many species of willow aphid, and they all like to eat! The beautiful willow tree seems to attract ravenous feeders. Look for wilting leaves and premature leaf drop. They can be a real nuisance if they fall off the tree. If you step on them, and that's hard to avoid doing, they leave red blood-like spots on the ground. If your tree is close to the house, they may make it inside, and those sticky spots are a real pain to clean up. Control is difficult without power spraying equipment.

Willow Borer (see, Tree Borers)

Yew

Black Vine Weevil

If this creature moves in, your plantings may practically disappear overnight. It chews on the roots and eats the center foliage—a one-two combination that is very effective. The larva is small and white, feeding on roots in the soil, and the adult weevil, which eats the foliage, is brown or black with yellow hairs.

Fletcher Scale (see, Scales)

Taxus Mealybug

If you spot a slow-moving, white insect that looks like someone sprinkled it with powdered sugar, it's probably a taxus mealybug. It whiles away its life by sucking plant juices.

Termites

Termites in your yew tree may spell trouble for your house. Call an exterminator fast! These tiny white creatures can be found on the branches, munching the bark and cambium. Look for dead branches.

Fruit Trees and Their Enemies

Fruit trees seldom die from disease. There are, of course, exceptions to this generalization. Leaf spots on sour cherry trees, black knot disease on plum trees, and fireblight disease on some varieties of apple and pear trees are serious threats. In general, however, disease usually weakens a tree, lowering the quality and quantity of the crop, rather than killing it.

Most fruit trees die because of injury to the trunk, crown (the area at ground level), and roots. Damage may be the result of severe winters, drought, excessively wet soil, or machinery such as lawn mowers, which sometimes strip off bark. Rabbits or small rodents can damage the tree by their nibbling. Weed-killers applied to grass underneath the tree can poison it, as the herbicide enters the root system.

Perhaps the major cause of damage, however, is insects, particularly borers. Without a good pest control program, you will never produce the kind of fruit you see in markets. All-purpose fruit sprays are readily available at garden centers and hardware stores. They generally contain a broad-spectrum insecticide and fungicide combination. The spraying program usually lasts as long as the growing season.

The Apple Maggot

The apple maggot (Color Figure 45) is representative of the pests that love your fruit trees as much as you do. Have you ever bitten into an apple and noticed brown streaks on the inside? Are your apples lumpy or undersized? These are indications of apple maggots.

The adult is a fly about one-fourth inch long. The body is black with white bands on the abdomen and dark bands on each wing. As they walk over the leaves and fruit, they constantly taste the surface. After mating, the females look for fruit in which to deposit their eggs. They lay there just under the skin of the fruit. Five to ten days later, the larvae hatch and tunnel through the flesh. Survival rates are low in hard, green fruit, but increase as the fruit ripens. The larvae usually mature after the fruit has fallen from the tree. The mature larvae leave the fruit and tunnel two or three inches into the soil. They spend the winter underground in cocoons, emerging as adult flies between early June and early September.

Apple Maggot Damage

There are two possible methods of control: (1) disposing of fallen fruit before the larvae emerge and (2) eliminating the adult fly with insecticide sprays.

The first method is practical for the average homeowner who has only a few trees to worry about. The apples should be destroyed before the larvae are mature enough to leave the fruit.

The second method is used in commercial fruit production and is definitely better if a large number of trees must be protected. The feeding habits of the flies make them vulnerable to stomach and contact poisons. Good coverage of the entire tree is essential, and spraying other trees and bushes in the vicinity of the fruit tree is usually desirable.

Spraying Techniques

Whenever you decide to use pesticides, certain precautions are absolutely essential. Check the charts for recommended sprays and amounts to use. The Cooperative Extension Service office in your county will be glad to advise you about the timing of applications (see list of CES addresses). Read the chapter on pesticides and sprayers before you begin. Be sure to read the instructions on the pesticide label and follow them carefully.

SPRAYING GUIDE FOR FRUIT TREES

Height of Tree (feet)	Branch Spread (feet)	Gallons of Diluted Spray per Tree per Application
4	3	0.5
5 to 8	3 to 6	1
8 to 10	4 to 8	2
10 to 15	8 to 15	3 to 6
15 to 20	15 to 25	7 to 10

SPRAYING GUIDE FOR TREES AND SHRUBS

Pesticides*	Amount Needed for Volume of Spray	
	One Gallon	Six Gallons
Bacillus thuringiensis	2 teaspoons	3/4 cup
Carbaryl	2 teaspoons	4 tablespoons
Chlorpyrifos	2 teaspoons	4 tablespoons
Diazinon	2 teaspoons	4 tablespoons
Dicofol	2 teaspoons	4 tablespoons
Dimethoate	2 teaspoons	4 tablespoons
Malathion	2 teaspoons	4 tablespoons
Methoxychlor	2 teaspoons	4 tablespoons

* *Use only the pesticides listed. Read label directions carefully before using any spray and follow directions exactly.*

Cooperative Extension Service

Experienced and inexperienced gardeners alike sometimes run into problems they cannot solve by themselves, even with the help of the local garden center and friendly neighbors. At these times, one of the most valuable bits of information any gardener can have on hand is how to get in touch with the local Cooperative Extension Service.

The Cooperative Extension Service was created by the Smith-Lever Act (1914) as an organizational entity of the United States Department of Agriculture and the state land-grant colleges and universities. Its purpose is to disseminate information from experts in various areas to the public through publications, correspondence, and other educational activities of an informal, problem-oriented nature. These activities are carried out primarily by the extension staffs at the county and state levels.

State Extension Service Addresses

Listed here are the names and addresses of the land-grant colleges and universities that serve as state headquarters for the Cooperative Extension Service. For general information regarding the overall services available in your state, or to find out the address of your local extension service, direct your correspondence to the Extension Director of your state's land-grant institution. For more specific questions, write to the extension expert in the appropriate field: Extension Horticulturist (cultural problems, recommended plant varieties, availability); Extension Entomologist (bugs and insects); or Extension Plant Pathologist (diseases).

Example: Appropriate Title
College of Agriculture
State Land-Grant Institution
City, State, Zip Code

Alabama: Auburn University, Box 95151, Auburn, AL 36830

Alaska: University of Alaska, Fairbanks, AK 99701

Arizona: University of Arizona, Tucson, AZ 85721

Arkansas: University of Arkansas, Fayetteville, AR 72701

California: University of California, Berkeley, CA 94720

Colorado: Colorado State University, Fort Collins, CO 80521

Connecticut: University of Connecticut, Storrs, CT 06268

Delaware: University of Delaware, Newark, DE 19711

District of Columbia: Federal City College, Washington, DC 20001

Florida: University of Florida, Gainesville, FL 32601

Georgia: University of Georgia, Athens, GA 30601

Hawaii: University of Hawaii, Honolulu, HI 96822

Idaho: University of Idaho, Moscow, ID 83843

Illinois: University of Illinois, Urbana, IL 61801

Indiana: Purdue University, W. Lafayette, IN 47907

Iowa: Iowa State University, Ames, IA 50010

Kansas: Kansas State University, Manhattan, KS 66502

Kentucky: University of Kentucky, Lexington, KY 40506

Louisiana: Louisiana State University, Baton Rouge, LA 70803

Maine: University of Maine, Orono, ME 04473

Maryland: University of Maryland, College Park, MD 20742

Massachusetts: University of Massachusetts, Amherst, MA 01002

Michigan: Michigan State University, East Lansing, MI 48823

Minnesota: University of Minnesota, St. Paul, MN 55101

Mississippi: Mississippi State University, Mississippi State, MS 39762

Missouri: University of Missouri, Columbia, MO 65201

Montana: Montana State University, Bozeman, MT 59715

Nebraska: University of Nebraska, Lincoln, NB 68503

Nevada: University of Nevada, Reno, NV 89507

New Hampshire: University of New Hampshire, Durham, NH 03824

New Jersey: Rutgers, The State University, New Brunswick, NJ 08903

New Mexico: New Mexico State University, Las Cruces, NM 88003

New York: Cornell University, Ithaca, NY 14850

North Carolina: North Carolina State University, Raleigh, NC 27607

North Dakota: North Dakota State University, Fargo, ND 58103

Ohio: The Ohio State University, Columbus, OH 43210

Oklahoma: Oklahoma State University, Stillwater, OK 74074

Oregon: Oregon State University, Corvallis, OR 97331

Pennsylvania: Pennsylvania State University, University Park, PA 16802

Puerto Rico: University of Puerto Rico, Rio Piedras, PR 00927

Rhode Island: University of Rhode Island, Kingston, RI 02881

South Carolina: Clemson University, Clemson, SC 29631

South Dakota: South Dakota State University, Brookings, SD 57007

Tennessee: University of Tennessee, P.O. Box 1071, Knoxville, TN 37901

Texas: Texas A & M University, College Station, TX 77843

Utah: Utah State University, Logan, UT 84321

Vermont: University of Vermont, Burlington, VT 05401

Virgin Islands: Virgin Islands Agriculture Project, Kingshill, St. Croix, VI 00801

Virginia: Virginia Polytechnic Institute and State University, Blacksburg, VA 24061

Washington: Washington State University Pullman, WA 99163

West Virginia: West Virginia University, Morgantown, WV 26506

Wisconsin: University of Wisconsin, Madison, WI 53706

Wyoming: University of Wyoming, Laramie, WY 82070

Tree and Shrub Pests

CONTROL SUGGESTIONS FOR TREE AND SHRUB PESTS

Pest	Sprays	Comments
Ailanthus webworm	*Bacillus thuringiensis.* carbaryl, Diazinon, malathion	Treat only if larvae are numerous
Aphids		See specific listings
Apple maggot	All-purpose fruit spray	Spray once before tree blooms; repeat every 10 to 14 days until late August
Arborvitae leafminer	Diazinon, malathion	Check with your CES*
Ash borer	Chlorpyrifos	Check with your CES*
Bagworm	Acephate	Check with your CES*
Banded elm leafhopper	Carbaryl	Treat if insects are numerous
Birch leafminer	Dimethoate	Treat when mines first appear in spring
Birch skeletonizer	Carbaryl, Diazinon, malathion	Treat when insects are first noticed
Black vine weevil	Diazinon	Check with your CES*
Black walnut curculio	Control unknown	————
Borers		See specific listings
Boxelder aphid	Diazinon, malathion	Treat only if a nuisance or if damage is severe
Boxelder bug	Carbaryl, Diazinon	Treat only if a nuisance or if damage is severe
Boxelder leafminer	Diazinon	Check with your CES*
Boxwood leafminer	Diazinon, dimethoate	Apply two to four weeks after new growth begins; check with your CES*
Bronze birch borer	Dimethoate	Check with your CES*
Brown-tail moth	Control unknown	————
Carpenterworm	Control unknown	————
Catalpa sphinx	Acephate, carbaryl, malathion	Treat when insects are numerous
Comstock mealybug	Acephate, malathion	Treat when insects are numerous
Cooley spruce gall aphid	Diazinon, malathion	Check with your CES*
Cottonwood borer	Control unknown	————
Cottonwood leaf beetle	Carbaryl, Diazinon, malathion	Treat when insects are numerous
Cottony maple leaf scale	Diazinon, malathion	Check with your CES*
Cottony maple scale	Diazinon, malathion	Check with your CES*
Cynthia moth	Control unknown	————
Dogwood borer	Chlorpyrifos	Check with your CES*
Dogwood scale	Diazinon, malathion	Check with your CES*
Eastern spruce gall aphid	Carbaryl, chlorpyrifos, Diazinon, malathion	Check with your CES*
Eastern tent caterpillar	Acephate, carbaryl, chlorpyrifos, malathion	Treat when tents appear; prune out tented areas
Elm borer	Control unknown	————

**Cooperative Extension Service*

Tree and Shrub Pests

CONTROL SUGGESTIONS FOR TREE AND SHRUB PESTS

Pest	Sprays	Comments
Elm cockscomb gall aphid	———	No control needed
Elm leaf aphid	Diazinon, malathion	Treat when insects are numerous
Elm leaf beetle	Acephate, carbaryl, Diazinon, malathion	Treat when insects are numerous
Elm spanworm	*Bacillus thuringiensis,* carbaryl Diazinon, malathion	Treat when insects are numerous
Eriophyid mite	Dicofol	Control usually is not necessary
Euonymous scale	Diazinon, malathion	Check with your CES*
European elm scale	Malathion	Treat in early spring when the leaves are forming
European pine shoot moth	Dimethoate	Check with your CES*
European red mite	Dicofol	Treat in early spring when the leaves are forming
Fall webworm	Acephate, *Bacillus thuringiensis,* carbaryl chlorpyrifos, Diazinon, malathion	Treat when insects are numerous
Flat-headed apple tree borer	Dimethoate	Check with your CES*
Fletcher scale	Malathion	Check with your CES*
Forest tent caterpillar	*Bacillus thuringiensis,* carbaryl, chlorpyrifos	Spray when the caterpillars first hatch in early spring
Galls	———	See specific listings
Gouty oak gall	Control unknown	———
Green-striped maple worm	———	No control needed
Gypsy moth	Acephate, *Bacillus thuringiensis,* carbaryl	Treat when larvae are small
Hackberry nipple gall psyllid	Diazinon, malathion	Check with your CES*
Hawthorn lace bug	Diazinon, malathion	Treat when insects are numerous
Hawthorn leafminer	Diazinon, malathion	Treat when mines first appear in spring
Hawthorn mealybug	Diazinon, malathion	Treat when insects are numerous
Hickory bark beetle	Control unknown	———
Hickory gall phylloxera	Malathion	Treat when leaves begin to break
Hickory horned devil	———	No control needed
Holly leafminer	Malathion	Treat when mines first appear in spring
Honey locust pod gall	———	No control needed
Japanese beetle	Carbaryl, Diazinon, malathion	Treat when insects are numerous
June bug	Carbaryl	Treat when insects are numerous
Juniper bark beetle	Control unknown	———
Juniper scale	Diazinon, malathion	Check with your CES*

**Cooperative Extension Service*

Tree and Shrub Pests

CONTROL SUGGESTIONS FOR TREE AND SHRUB PESTS

Pest	**Sprays**	**Comments**
Juniper webworm	Diazinon, malathion	Check with your CES*
Leafhoppers	———————	See specific listings
Leafminers	———————	See specific listings
Lilac borer	Chlorpyrifos	Check with your CES*
Locust borer	Carbaryl	Check with your CES*
Locust mite	Dicofol	Treat in early spring when the leaves are small; repeat every ten days as needed
Magnolia scale	Carbaryl, dormant oils, malathion	Check with your CES*
Maple bladder-gall mite	———————	No control needed
May beetle	Carbaryl	Treat when insects are numerous
Mealybugs	Acephate, Diazinon, malathion	Treat when insects are numerous
Mimosa webworm	Acephate, *Bacillus thuringiensis,* carbaryl chlorpyrifos, Diazinon, malathion	Treat when webs appear
Nantucket pine moth	Acephate, dimethoate	Check with your CES*
Norway maple aphid	Diazinon, malathion	Treat when insects are numerous
Oak apple gall	———————	No control needed
Oak borer	Control unknown	———————
Oak kermes	Diazinon, malathion	Check with your CES*
Oak leafminer	Diazinon, malathion	Check with your CES*
Oystershell scale	Diazinon, malathion	Check with your CES*
Painted maple aphid	Diazinon, malathion	Treat when insects are numerous
Periodical cicada	Carbaryl	Treat when insects are numerous
Pigeon tremex	———————	No control needed
Pine bark aphid	Diazinon, malathion	Check with your CES*
Pine needle scale	Acephate, chlorpyrifos, Diazinon, malathion	Check with your CES*
Pine tortoise scale	Carbaryl	Check with your CES*
Pit-making oak scale	Carbaryl	Check with your CES*
Poplar borer	Control unknown	———————
Poplar tentmaker	Acephate, Diazinon, malathion	Treat when insects are numerous
Privet thrips	Acephate, malathion	Treat when insects are numerous or causing damage
Redbud leafhopper	Carbaryl	Treat when insects are numerous
Red-headed ash borer	Control unknown	———————
Red-headed sawfly	Acephate, Diazinon, malathion	Treat when insects are numerous

**Cooperative Extension Service*

Tree and Shrub Pests

CONTROL SUGGESTIONS FOR TREE AND SHRUB PESTS

Pest	Sprays	Comments
Regal moth	————	No control needed
San Jose scale	Diazinon, malathion	Check with your CES*
Scales	————	See specific listings
Scurfy scale	Diazinon, malathion	Check with your CES*
Smaller European elm bark beetle	Methoxychlor	Check with your CES*
Snow-white linden moth	*Bacillus thuringiensis*, carbaryl Diazinon, malathion	Treat when insects are numerous
Southern red mite	Dicofol	Treat when leaves are forming in spring
Spirea aphid	Acephate, Diazinon, malathion	Treat when insects are numerous
Spring cankerworm	Acephate, *Bacillus thuringiensis*, carbaryl	Check with your CES*
Spruce bud scale	Diazinon, malathion	Check with your CES*
Spruce budworm	Diazinon, malathion	Check with your CES*
Spruce spider mite	Dicofol	Treat when insects are numerous and causing damage
Sweetgum scale	————	No control needed
Sycamore borer	Chlorpyrifos, dimethoate	Check with your CES*
Sycamore lace bug	Diazinon, malathion	Treat when insects are numerous
Taxus mealybug	Acephate, Diazinon, malathion	Check with your CES*
Termite	Diazinon	Treat if insects are causing damage
Tulip tree aphid	Diazinon, malathion	Check with your CES*
Tulip tree scale	Dormant oil	Check with your CES*
Twig pruner	Control unknown	————
Two-spotted spider mite	Dicofol	Check with your CES*
Walnut caterpillar	Acephate, carbaryl, malathion	Treat when larvae are numerous
White-marked tussock moth	Carbaryl, Diazinon, malathion	Treat when larvae are numerous
White pine weevil	Control unknown	————
Willow aphid	Diazinon, malathion	Treat when aphids are numerous
Willow borer	Control unknown	————
Winged euonymous scale	Diazinon, malathion	Check with your CES*
Witches'-broom	Control unknown	————
Wooly elm aphid	Diazinon, malathion	Treat when insects are numerous
Wooly hawthorn aphid	Diazinon, malathion	Treat when insects are numerous
Wool-sower gall	————	No control needed
Yellow-necked caterpillar	Acephate, carbaryl, malathion	Treat when insects are numerous
Zimmerman pine moth	Dimethoate	Check with your CES*

**Cooperative Extension Service*

MINI-ACRES
COUNTRY
CLUB

Lawn Pests

Lawn insects live under the soil, on the ground, and on the grass itself. Preventive measures can keep your lawn rich and green — the way you like it.

A RICH green lawn enhances the beauty and the value of any home. A thick carpet of grass, however, does require effort on your part. There are many insects and insect-like pests that damage or destroy your landscaping, leaving large areas of your lawn ragged, discolored, or even dead.

Lawn-loving pests fall into several basic groups. Some live under the soil, feeding on plant roots; some scurry about the surface, munching grass blades and stems; still others flit from plant to plant, sucking out the juices. There are also insects and small animals that damage your lawn indirectly by constructing dirt mounds that constrict and choke the grass. In addition, there are a number of insects that just like living in the lawn. They do little damage to the plants, but are considered pests because they bother people and their pets.

Lawn pests are usually not difficult to eliminate. The chart at the end of this chapter will give you suggestions about how to handle the problems you've spotted. Before beginning treatment, however, you must identify the pest. We have divided this descriptive list of major pests into two groups—insects living beneath the surface and insects living above ground.

What's the Problem Down There?

Ants

Ants are probably the world's number one pest. They can be found almost anywhere. Most species, especially those classified as turf headaches, build their nests in the ground, forming little hills near the nest entrances. These ant hills and mounds are ugly and very often smother the surrounding grass. Ants are capable of damaging the roots of grass plants by nesting in and around them. They have also been known to destroy the seeds, thus preventing full growth.

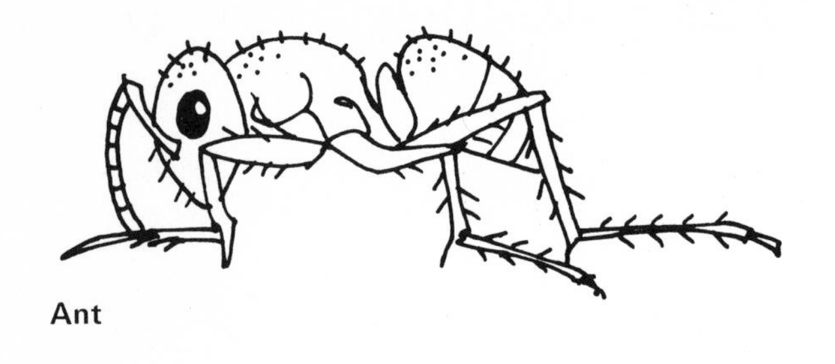
Ant

Control is relatively easy—all you have to do is spot drench the soil where the mounds are located.

Bees and Wasps

Cicada-Killer Wasps—Once you see this baby, you won't forget it! This wasp is huge, about two inches long, and a fast, strong flier. The body is yellow with black stripes. The cicada-killer wasp works like an undertaker. It digs grave-like nests in the ground. This crypt is prepared for the periodical cicada on which the wasp preys. The female wasp stings the cicada, paralyzing and embalming it at the same time. Once the cicada's body is placed in its tomb, an egg is deposited on it. A few days later, the egg hatches and the newly emerged larva feeds on the captured cicada. Sounds like a great late-late show horror movie, doesn't it?

Cicada-Killer Wasp

These wasps are expert diggers, throwing up small mounds of soil at the entrances to the nests. The holes are three-fourths inch in diameter and extend six inches or so from the entrances. They usually don't sting people, but when they do—ouch! Control is only necessary if the dirt mounds are unsightly or if the wasps are a nuisance. By killing destructive cicadas, they may actually be helpful.

Soil-Nesting Bees and Wasps—There are many different kinds of soil-nesting bees and wasps. They feed on a variety of plants and other insects that inhabit the area. They usually aren't much of a problem, but they should be destroyed if they become nuisances.

Control measures consist of drenching the soil with the proper insecticide. Work in the evening hours when the activity level of the insects is lowest. After treatment, cover the nest with a shovelful of dirt.

Billbugs

The billbug resembles the white grub but is legless. A case of mistaken identity is only likely in the larval stage. The adult is a dark-colored weevil or snout beetle. (By the way, all weevils are beetles. but not all beetles are weevils. Weevils are beetles with snouts, looking a bit like miniature elephants.)

There are different species of billbug around the country. A typical life cycle begins in the spring when the female lays her eggs in the stems of your grass. Tiny

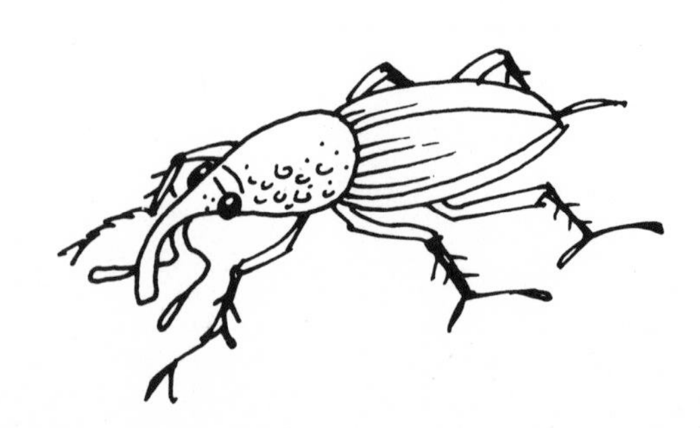
Billbug

larvae hatch and begin feeding on your grass stems, working their way right down into the root zone. They chomp away all summer until they pupate in autumn.

Look for browning in small patches. If you don't see any adult beetles, check a piece of grass. The stem will have a fine, powdery material where it has been chewed.

Control measures are not always necessary. Even though you will sometimes spot both adult and larvae billbugs, they may not be present in large enough numbers to do much damage.

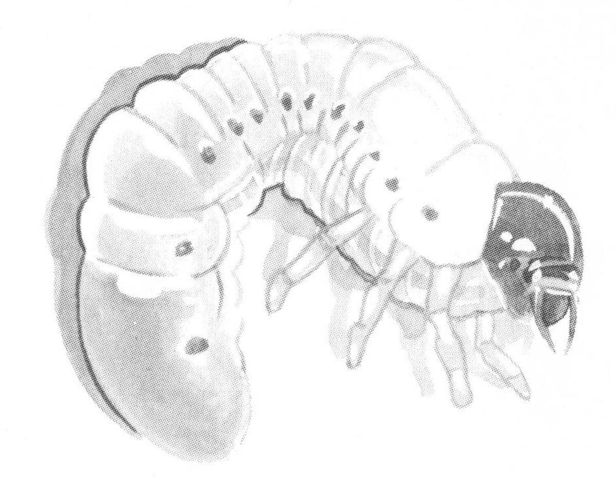

White Grub

Ground Pearls

These pearls aren't worth a dime, so forget your dreams of buried treasure! The hard, round shells that resemble tiny pearls (about one-eighth inch in diameter) enclose immature scale insects. The young insects feed on the fine rootlets of Bermuda and centipede grass by sucking out the plant juices. When ground pearl infestation is heavy, the grass turns brown and dies in irregular patches.

Chemical control doesn't work, but good lawn care, especially fertilization and irrigation, will help minimize the damage.

Grubs

The white grub, as it is commonly called is the larval form of a number of different beetles. The grub worm has a U-shaped body and a brown head. The body is white with a transparent abdomen—that's right, you can faintly see the body contents through the skin (Color Figure 58).

The adult beetle lays its eggs in early summer. When the larvae hatch, they burrow into the root zone of your grass and eat and eat and eat. As the cold weather approaches, the grub worm burrows deeper into the soil. In northerly climates, they go below the frost line. As spring nears and the ground thaws, they return to the surface. Depending on the species, grubs will spend one to three summers eating away before maturing into adult beetles. The adult varieties include the notorious Japanese beetle (Color Figure 60), the European chafer, the green June beetle, the white-fringed beetle, the Oriental beetle, the Asiatic garden beetle, and the annual white grub. Some of these adults feed on the foliage of shrubs and trees. Severe infestations may cause extensive leaf damage. Oak trees are particularly vulnerable (Color Figure 62).

The voracious appetites of these grub worms will turn large, irregular patches of your lawn brown, soft, and spongy. Sometimes the grass can be picked up and rolled back just like a carpet. If you're lucky, you may discover the grubs early in the year when you decide to turn over some sod for a garden. Put two and two

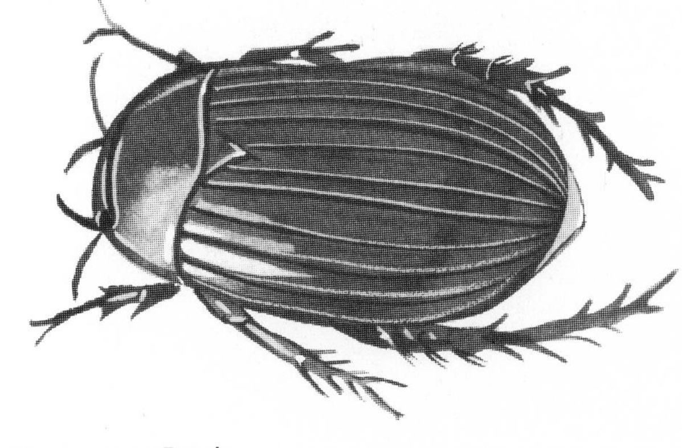

May or June Beetle

together fast, and start your treatment before they have a chance to really get going. In most cases, however, you won't know you have a problem until your grass turns brown. This is most likely to occur in the fall when the grubs are larger and harder to control. Be patient and keep calm. Jumping up and down on the grass will not help at all. Be sure your insecticide soaks well into the root zone where the grubs are feeding. Don't forget to use plenty of water.

Just to make things even more delightful, skunks, racoons, moles, and other small animals have been known to visit grub-infested lawns in search of these delectable gourmet snacks. Many homeowners have returned from vacation to find lawns that look like the site of the D-Day invasion. By removing the grubs, you will also eliminate this secondary animal problem.

Mole Crickets

Shovel-like feet and stout forelegs give this burrowing little creature its very appropriate name. Mole crickets love to munch on grass roots. They also make their presence known by uprooting entire grass plants, causing the soil to dry out quickly. One mole cricket can create havoc in a square yard of turf in a single night. They are usually seen in the south Atlantic and Gulf Coast states, but have been found in other areas as well.

The mole cricket usually doesn't require control measures. The keyword here is "usually." On occasion they get out of control and you'll have to get busy.

Mole Cricket

Cicada

Periodical Cicadas

Most of the basic information about these frog-like insects can be found in the chapter "Tree and Shrub Pests." Here we'll just briefly discuss the damage they do to lawns. Although the adult cicada (Color Figure 43) doesn't eat, it tunnels through the soil to the surface as soon as it reaches maturity. A large population can make your lawn look like a Swiss cheese.

It is virtually impossible to control the insect in its soil-dwelling stage because it lives pretty far below the surface for years. Control is directed to the adult insects in the trees in an attempt to reduce egg-laying and, therefore, population levels.

Wireworms

These insects are not attracted to wires nor do they feed on them. They get their name from the fact that they resemble thin, rust-colored wires. They are shiny little creatures about one inch long with a hard covering (Color Figure 66). Depending on the species, they live in the soil for two to six years feeding on grass roots. The adult wireworm is known as the click beetle. It will lie on its back and play dead, but will flip up into the air when approached too closely.

When the wireworm population is heavy, you'll notice irregular patches of dead brown grass around your yard. If you check the edges of these patches you may see the wireworms eating away.

Wireworm Larva

Surface Problems Are Not Superficial

Aphids

Aphids (Color Figure 73) seem to create problems almost everywhere. That's the sign of a well-rounded pest—a Renaissance insect, so to speak. These creatures have already been discussed in the chapter "Tree and Shrub Pests," but they are worth mentioning again. Some aphids congregate on grass leaves and do a great Dracula imitation—they suck the life right out of them. Look for yellowing leaves and rounded brown areas.

Everybody wants to get into the act when it comes to eliminating these unpleasant creatures. Ladybugs are their natural enemies, feeding on them most happily, and fungus diseases are effective against them in warm, moist weather. Perhaps man is the aphid's worst enemy—insecticides can be very effective.

Armyworms and Cutworms

Armyworms and cutworms get their name from the fact that groups of them invade an area and level it with marvelous discipline before moving on. They chew away, usually in circular patches (Color Figure 63).

Control measures are seldom necessary because fungus diseases and natural predators will take care of this problem for you.

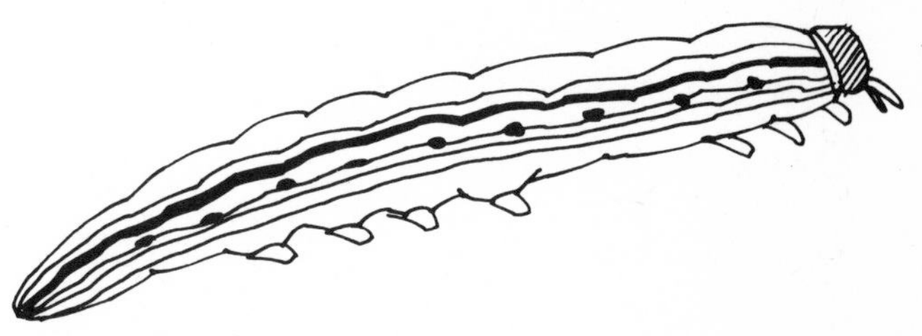

Armyworm Larva

Chinch Bugs

Chinch bugs are sun-lovers. The scattered yellow patches of grass that result from their activities are usually seen in the sunny areas of the lawn. The greatest damage appears in July and August, when the immature insect is sucking the plant juices.

A good way of testing for possible chinch bug infestation is to take an old coffee can (if you're rich enough to afford one) and remove the top and bottom, leaving a hollow tube. Shove the can into the turf inside the yellowed area and then fill it with water. Look for small white and black insects about 1/16 inch long floating on the surface. Be patient, because it will take about ten minutes for them to float to

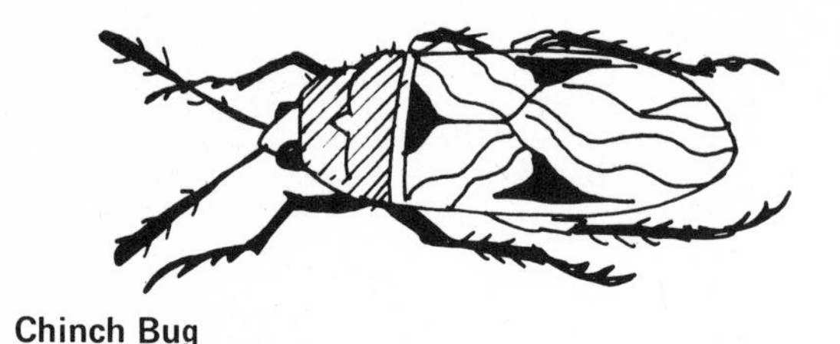

Chinch Bug

the top. Once you've identified the problem, get rid of it!

Crane Flies

These insects look like giant mosquitoes. The adults do not damage grass, but the larvae feed on the roots, occasionally causing brown patches. Since this happens so seldom, controls are really unnecessary.

Crickets

Dismiss your fond memories of Disney movies and cute television commercials. Crickets can be real pests! If your house happens to be in their path, they will enter through cracks and crevices or under doors and spend hours serenading you.

As long as they stay outside you don't have to worry. They won't damage the lawn. Once they start showing up inside, however, control measures may be necessary.

Earthworms

A few species of earthworms can cause turf damage by throwing up substantial amounts of soil, thus smothering the grass. For the most part, however, they are beneficial insects because they aerate heavy soils.

Earwigs

These insects got their name from an old superstition. People used to believe they would crawl into the ear of a sleeping human, causing all sorts of difficulties! In fact, however, earwigs are completely harmless. They are dark brown, one-half to three-fourths inch long, and have a pair of formidable-looking pinchers at the end of their tails (Color Figure 17). They come out at night to feed. They like almost any kind of plant, but their very favorite dish is a pile of decaying organic material such as grass clippings.

You can safely ignore them as long as they stay outside, but take action fast if they cross your threshold. Once they're inside, they hide in cracks and their habits will remind you of the unpleasant cockroach.

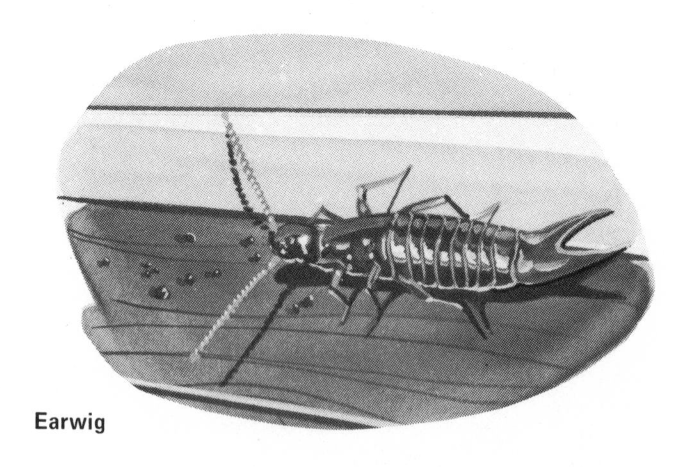

Earwig

Fleas

Fleas are an occasional lawn denizen. The adult female lays her eggs wherever she travels. Once the eggs hatch, the young fleas start looking for someone to feed on. You or your pet are likely victims (Color Figure 15). Sprays are effective in controlling infestation. Pets should be dusted with flea and tick powder.

Frit Flies

These characters have been getting a lot of publicity lately, mainly because they hang around golf courses. The adults are small black flies, but the damage is done by the immature maggots that tunnel into grass stems near the surface of the soil. This tunneling causes the upper portion of the plant to die. On golf courses, the damage first appears around the periphery and gradually moves inward. Bluegrasses are particularly susceptible. Positive identification requires laboratory study of the larvae, but if you find a golf ball covered with black flies, you should suspect these insects.

Grasshoppers

Grasshoppers (Color Figure 68) are nuisances rather than pests. When the population is heavy, it is sometimes impossible to walk across a lawn without disturbing them. A large number of them can cause considerable destruction to grass leaves and stems.

Leafhoppers

Leafhoppers are small bullet-shaped insects that seem to hop from leaf to leaf, sucking plant juices. Actually, they do not hop; they simply fly short distances. They come in a variety of colors. When the population is heavy, your lawn will become a lighter shade in certain areas. See the chapter "Tree and Shrub Pests" for more details. Control is seldom necessary, but a very heavy population might kill a newly planted lawn.

Lucerne Moths

In all likelihood, you wlll never have problems with these insects, but, on the other hand, don't underestimate bugs. The larvae of this species feed primarily on the clover in your lawn. They resemble sod webworm larvae, but are a little bigger. Look for them in late summer.

Millipedes and Centipedes

These ugly little creatures are not insects,

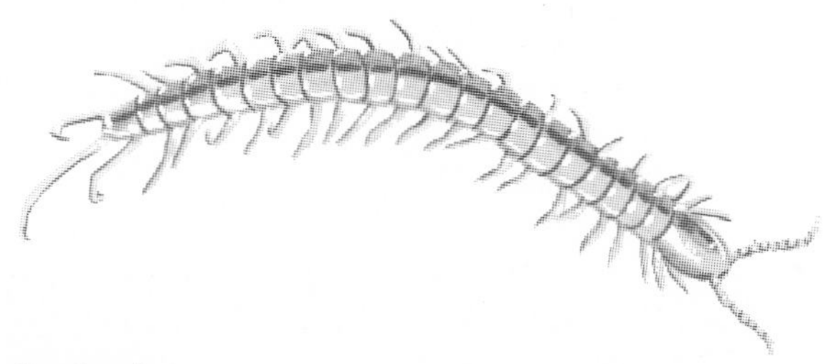

Centipede

Millipedes

but tiny worm-like animals with many body segments and legs. The difference between the two is that centipedes (100-leggers) have two legs per body segment while millipedes (1000-leggers) have four legs per body segment (Color Figures 38 and 39). Both species curl up into a coil when disturbed.

They usually don't damage lawns because they feed mainly on decaying matter. They can be a real nuisance, however, when they congregate in great numbers after a heavy rain. In spring and fall you may see multitudes of them crawling through the grass as they migrate. If your house happens to be in their path, watch out! They crawl into houses, garages, pools, and any other structure in their way (see "People Pests").

Mites

Mites are tiny creatures more closely related to spiders and ticks than to insects. There are several different species within the United States, feeding on a variety of lawn plants.

The damage they cause is usually not severe, but they can become quite pesty if they find their way inside as they sometimes do in early spring and fall (see

"People Pests"). Chemical controls are not necessary unless the many natural predators that feed on mites are falling down on the job.

Mosquitoes

These blood-suckers (Color Figure 14) are an aggravating nuisance. Almost everyone has been bitten by them at some time. They can be found anywhere and everywhere during summer months, and seem to be particularly nasty in the evening hours. Although mosquito bites are minor problems for most people, a few individuals experience severe allergic reactions and they can present serious public health problems. See "People Pests" for further information.

Sandflies

These awful pests are sometimes called "no-seeums" or "punkies" because they bite and are gone almost before you can react. This group of midges or flies are annoying and their bites can really hurt!

Scale Insects

There are many species of scale insects. They attack not only lawns but other plants as well (see "Tree and Shrub Pests"). Scales are tiny growths that are often overlooked because they appear to be part of the plant itself. The eggs are deposited under these growths. The newly hatched insects are crawlers that eventually settle down on a leaf or stem and begin sucking the plant juices. At this time, the crawlers cover themselves with protective shells. Two species of scales—the Rhodesgrass scale and the Bermudagrass scale—are especially fond of lawns. Populations must be heavy, however, before any severe damage will result.

If controls are necessary, you won't find an easy way out. Scales can only be controlled in the crawling stage, because insecticides cannot penetrate their protective shells. Repeated applications may be required.

Sod Webworm Larva

Sod Webworms (Lawn Moths)

These moths, like other mature moths, don't feed; they only reproduce. The damage is done by the larvae who love to eat. Perhaps they suspect that each meal may be their last.

Sod webworm damage is easily spotted (Color Figure 61). Feeding takes place near the crown of the plant, causing irregular patches of brown grass. If the larval population is high, your lawn may become ragged within a few days. Look for small (one-inch wingspans), buff-colored moths that are attracted to lighted windows in the evening hours, and larvae or cocoons in grass thatches. The larvae are gray to very light brown in color with scattered dark spots (Color Figure 59). An unusual number of birds, especially robins, is a good indication of larval infestation.

If you think larvae are present, mix two or three teaspoons of detergent with a gallon of water and pour it over a square foot of turf. This will bring the larvae to the surface.

One nice thing about sod webworm damage is that the lawn will usually recover quickly. This is because the bases and root systems of the plants are not injured.

Sowbugs and Pillbugs

These pests are not insects, but they are nuisances (see "People Pests"). They look like armadillos, but one lady with a gift for words described them as miniature Volkswagens (Color Figure 41). Both species feed on organic material in the soil. Once in a blue moon, they may bother grass and other plants. While they seldom damage lawns, they have been known to run amuck in greenhouses.

Clear out decaying organic matter periodically, and they probably won't bother you. If they should get inside, foundation spraying should take care of the problem.

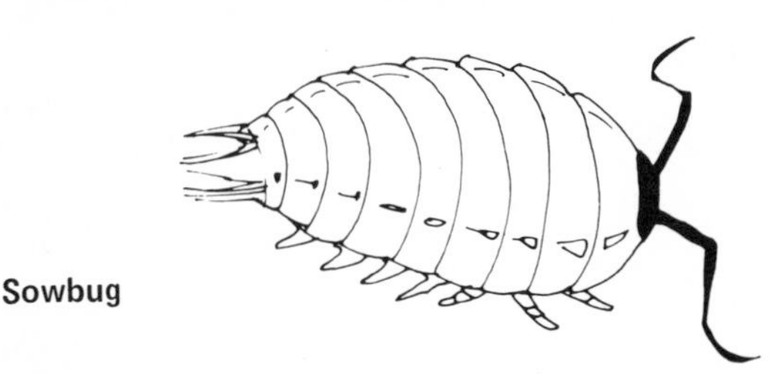

Sowbug

Spiders

These are probably the most feared of all the crawling creatures found near homes. Perhaps it's the way they look, or all those sci-fi horror movies we saw as kids showing giant spiders taking over the world. Despite the revulsion they arouse, spiders are almost always nuisances rather than threats (see "People Pests").

There are about 25,000 species found within the continental United States. They have been found on all sorts of shrubs and grasses and are often helpful in controlling insect populations. You may find them especially annoying in the fall when they look for a warm house to shelter them from the cold.

Spittlebugs

How did these insects get their silly name? These plant-suckers produce a substance that looks exactly like spit. The bugs, which look like leafhoppers, can often be seen in the middle of this froth (Color Figure 53). You won't forget this calling card!

Control measures are usually unnecessary because they cause only minor damage.

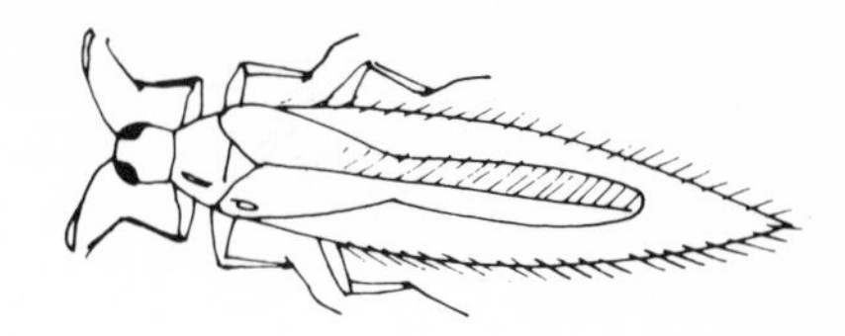

Thrips

Thrips

Thrips are the only insects that have rasping, sucking mouth parts. They scratch blades of grass and suck the plant juices. Your lawn will begin to look white or silver, a form of damage called silvertop. They feed on a variety of plants besides grass. Because they are so tiny (about one-eighth inch long), your best clue to their presence is the characteristic silvertop they produce.

Once in a while, one of these insects may crawl on your body and bite. The bites are annoying, but not poisonous. Control is relatively easy.

Ticks

Ticks, like fleas, like the blood of animals. Your pet will be their prime target. If your property is wooded or abuts a wooded area, you are much more likely to be bothered. Once your pet is infested, your house may be invaded by ticks carried inside by the unhappy animal. Take action fast! See "People Pests" for further information.

Spraying

You will never be able to eliminate lawn pests completely. Your goal should be to control them sufficiently to prevent damage to your lawn and excessive annoyance to your family.

Good lawn maintenance is your first line of defense. A well-watered and well-fertilized lawn is more able to resist insect attack and plant disease. Removing excess grass and shrubbery reduces possible hiding places for pests.

Once you have identified your particular problem, check the chart at the end of this chapter for control suggestions. Read the label directions carefully before using any means of chemical control. Remember that pesticides should be used only when necessary. Avoid spraying whenever possible.

In general, liquid concentrates are the easiest to use and they are also the longest lasting. These concentrates are mixed with water and applied as sprays. To control insects below ground, mix the insecticide with plenty of water (see label directions for exact proportions) and soak into the soil. Insecticides designed for insects living above the ground must remain in contact with the blades and stems to be effective. Read the chapter on pesticides for descriptions of spraying equipment and proper equipment care.

The spraying directions usually recommend a specific number of ounces of concentrate per a specific number of square feet of lawn surface. Proper application is crucial, so brush up on your basic math. Remember that area equals length times width. If your lawn is 40 feet long by 20 feet wide, the area is 800 square feet.

SUGGESTIONS FOR CONTROLLING LAWN PESTS

Pest	Pesticides*	Comments
Ants	Diazinon	Soak spray into soil, using plenty of water
Aphids	Diazinon, malathion	Apply spray to grass; do not soak into soil
Armyworms	Carbaryl, Diazinon	Spray grass when worms are seen feeding
Bees (ground-nesting)	Carbaryl, Diazinon	Soak spray into soil, using plenty of water; cover holes with dirt
Billbugs	Carbaryl, Diazinon	Soak spray into soil; repeat in 14 days
Centipedes	Carbaryl, Diazinon	Spray grassy areas near house; inside—use vacuum cleaner
Chiggers	Diazinon, malathion	Spray lawn; use a repellent containing Deet on yourself
Chinch bugs	Carbaryl, chlorpyrifos, Diazinon, propoxur	Water lawn before treating, using at least 25 gallons per 1000 square feet; repeat in 14 days
Cicada-killer wasps	Carbaryl, Diazinon	Soak spray into soil; using plenty of water; cover holes with dirt
Crane flies	————	No control needed
Crickets	Diazinon	Apply spray to grassy areas; do not soak into soil
Cutworms	Carbaryl, Diazinon	Apply spray to grass blades; do not soak into soil
Earthworms	————	No control needed
Earwigs	Carbaryl, chlorpyrifos, Diazinon	Spray grassy areas near house or areas where earwigs have been seen
Fleas	Carbaryl, malathion	Spray grassy areas where infestation is suspected; keep pets dusted with flea powder

**Use only one of the listed insecticides. Keep people and pets off the lawn until the spray has dried.*

Lawn Pests

Pest	Pesticides*	Comments
Frit flies	Diazinon	Spray areas where insects have been seen
Grasshoppers	Carbaryl	Apply spray directly to grass blades when insects are abundant
Ground pearls	Control unknown	————
Grubs	Chlorpyrifos, Diazinon	Soak spray into soil, using plenty of water, in summer or early fall
Leafhoppers	Carbaryl, Diazinon, malathion	Spray when insects are very abundant; grass blades must be thoroughly covered
Lucerne moth	Carbaryl	Spray only if damage is severe
Millipedes	Carbaryl, Diazinon	Spray grassy areas near house; inside — use vacuum cleaner
Mites	Diazinon, dicofol, malathion	Spray when mites are numerous; repeat in 14 days if insects or damage are seen
Mole crickets	Diazinon	Soak thoroughly into soil
Mosquitoes	Chlorpyrifos, malathion	Spray grassy areas and shrubs; use repellent containing Deet on yourself
Periodical cicada	Carbaryl	Spray trees and shrubs only when insects are very abundant
Pillbugs	Diazinon	Spray areas near buildings and areas of heavy infestation only when absolutely necessary
Sandflies	Malathion	Spray grassy areas and shrubs; use repellent containing Deet on yourself
Scales	Diazinon, malathion	Spray when scales are crawling
Sod webworms	Carbaryl, chlorpyrifos, Diazinon, propoxur	Spray when worms are active
Sowbugs	Diazinon	Spray areas near buildings and areas of heavy infestation only when absolutely necessary
Spiders	Diazinon	Spray foundations if spiders are entering the house
Spittlebugs	Carbaryl, Diazinon	Spray if insects or damage are seen
Thrips	Carbaryl, Diazinon	Spray when population is high
Ticks	Carbaryl, malathion	Spray grassy areas where infestation is suspected; keep pets dusted with tick powder
Wasps (ground-nesting)	Carbaryl, Diazinon	Soak spray into soil, using plenty of water; cover holes with dirt
Wireworms	Diazinon	Soak spray into soil, using plenty of water, in summer or early fall

**Use only one of the listed insecticides. Keep people and pets off the lawn until the spray has dried.*

Vegetable and Flower Pests

Blue-ribbon vegetables and breath-taking flowers attract insects as well as compliments. Why share your achievements with uninvited munchers?

MANY PEOPLE can't wait for the ground to thaw in spring because they're so anxious to get out in their gardens and start digging. Garden fever runs high in the beginning, but as warm weather approaches and the bugs make their entrance for the season, enthusiasm sometimes wanes. Controlling insects is one of the gardener's biggest headaches. More time and energy are expended on bug problems than on any other aspect of gardening.

What you must remember is that garden pests cannot be completely eliminated. If you grow snap beans, you will encounter bean beetles; if you are a rose fancier, you will always have to cope with aphids. Your goal is not to exterminate every insect in your garden—any measures strong enough to accomplish that may injure your plants and possibly harm you as well—but to keep the insect population under control and thus avoid serious damage to your crops.

Some gardeners give up completely when the bug hordes invade. Others treat their plants erratically, killing a few insects every now and then, and are surprised that they still find holes in their cabbage leaves. Still others—and these are the successful gardeners—weed regularly and carefully protect their plants against insects and disease by well-planned treatment. The members of this happy group usually run around the neighborhood distributing vegetables and cut flowers to their friends.

There's no reason why your garden can't be just as or even more successful. Just remember that trying to raise vegetables and flowers without protecting them from insects and disease is like trying to drive your car without gas. You won't get very far!

Identifying your particular pest is the first step. Some insects are very unselective eaters, but others concentrate on one type of plant. What you're growing can tell you what pests you are likely to find. Once you know the problem, you can begin to deal with it.

Most insects go through three or four stages of development—only one or two of which are destructive. Moths and butterflies, for example, are absolutely harmless as adults, but can be deadly to plants during their larval stage when they eat voraciously. Control measures must be directed to the proper developmental stage in order to be effective.

You must be on the alert from early spring until frost appears in autumn. Some insects spend the winter season in or close to the garden and are ready to go to work as soon as the temperature hits 45 or 50F. Other insects migrate to the garden later in the season.

The list describes some of the pests you are likely to encounter. The chart at the end of the chapter provides suggested measures to take if controls seem necessary.

The Pests

Aphids

These obnoxious creatures have already been discussed in earlier chapters, but they can't resist making still another appearance. There's probably a species of aphid for every plant in your garden and then some. They come in a variety of

TIME TO WAIT BETWEEN PESTICIDE APPLICATION AND VEGETABLE HARVESTING

Vegetable Crop	Days To Wait For		
	Carbaryl	Diazinon	Malathion
Beans	0	7	1
Cabbage and related crops	3	7	7
Collards and other leafy crops	14	—	7
Eggplant	0	—	3
Lettuce	14	—	14
Onions	—	—	3
Peas	0	0	3
Potatoes	0	—	0
Pumpkins	0	—	3
Tomatoes	0	3	1
Vine crops	0	3	1

Mature Winged Aphid

colors, ranging from pink to green to black, and any number of shades in between. Look for an awkward, soft-bodied insect that resembles a slow-moving pear (Color Figure 73). As the aphids suck the juices, the plant becomes stunted and does not produce a normal crop. Migrating aphids also transmit several important plant diseases.

Because aphids as a group live on such a variety of plants and have adverse effects in two ways—injury and disease—they must be considered the number one garden pest. Early control is important.

Bean Leaf Beetles

This little creature is a real ham actor—at the slightest disturbance it folds its legs, falls to the ground, and plays dead. It loves soybeans, green beans, cowpeas, and peanuts. Look for a green, yellow, tan, or red beetle with a distinctive black band around the outer edges of the wing covers feeding on the underside of the leaf.

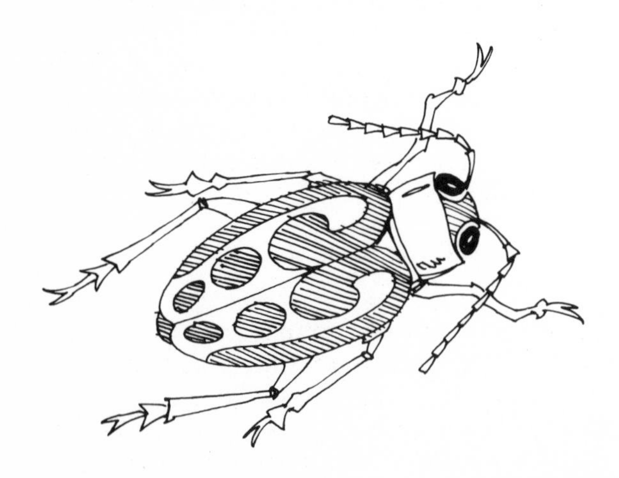

Bean Leaf Beetle

Blister Beetles

The blister beetle is a Jekyll-and-Hyde character. It is often beneficial because the newly hatched larva feeds on grasshopper eggs, effectively reducing the numbers of these pests. On the other hand, great swarms of beetles can create havoc in your garden. Striped blister beetles, for example, sometimes concentrate on specific rows of plants, virtually destroying the crop.

Look for them to appear in hordes during or just after a grasshopper outbreak. They are easily identified by the narrow area between the head and the wing covers that resembles a neck. Be careful if you pick one up—they contain an oily substance called cantharidin that can cause large blisters to form on your skin if you crush them.

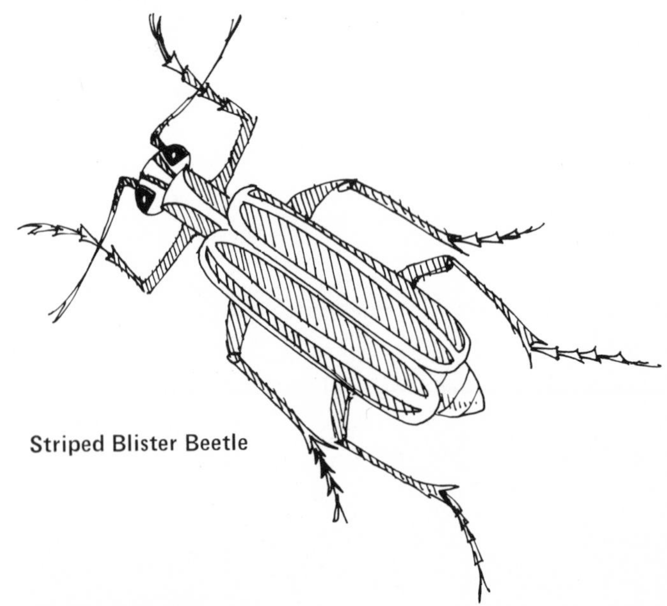

Striped Blister Beetle

Cabbage Worms

These common cabbage pests come in three major varieties: the imported cabbage worm, the cabbage looper (Color

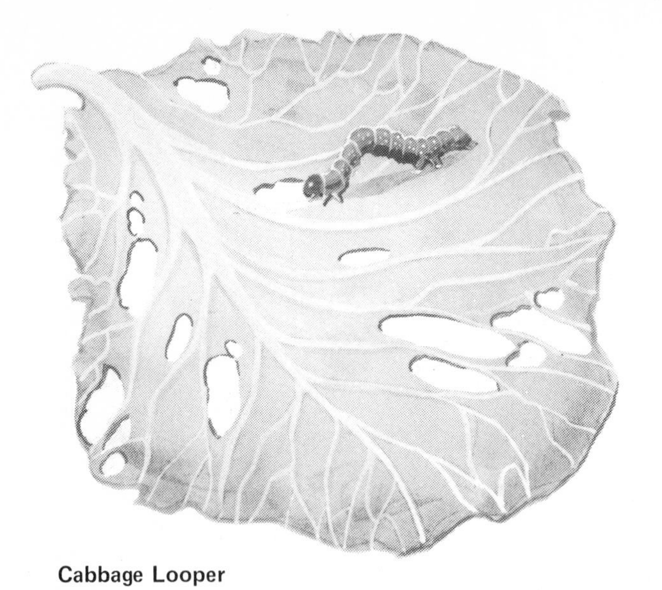

Cabbage Looper

Figure 70), and the diamondback moth larva. Each of them is capable of making your cabbage look like a Swiss cheese. They can completely defoliate a plant or eat their way deep into the cabbage heads. Once they're inside, control is extremely difficult.

Look for green worms ranging in size from 1¼ to 1½ inches long.

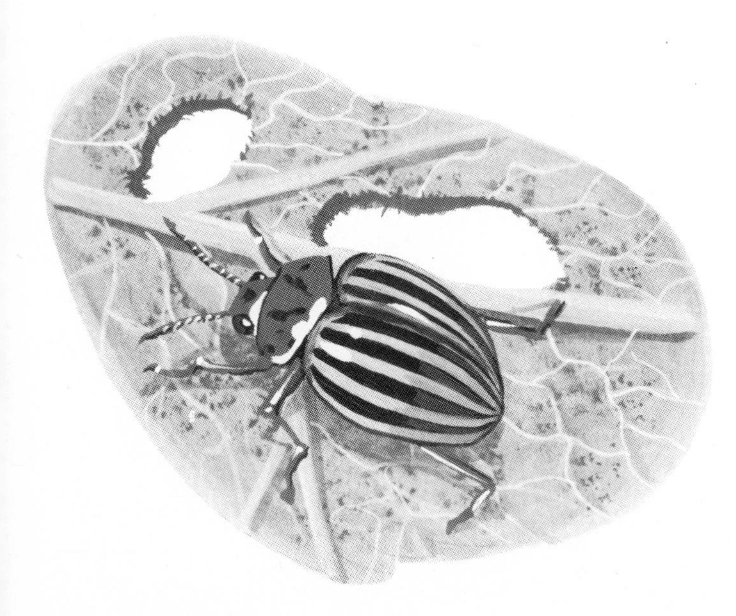

Potato Beetle

Colorado Potato Beetles

Perhaps the best known of American beetles, this insect shows up early in the spring. It prefers potatoes, but is almost as happy with tomatoes, eggplants, peppers, and some flowers.

You won't have any trouble recognizing this hungry insect. Look for a yellow, hard-shelled beetle with ten black lines on its back (Color Figure 71). You may also find orange-yellow egg clusters on the undersides of the leaves.

Corn Borers

The corn borer infects about 200 different kinds of plants, but it is particularly harmful to corn. It leaves bent or broken stalks as its calling card. Look for holes and sawdust on the stalks.

Good sanitation is important to control. Remove the infested stalks during the off-season. Treat the whorl and ear zones, especially if you grow early corn.

Corn Flea Beetles

This pest is something of an exception to the general rule in that it is the adult beetle rather than the larva which is harmful. The very small, shiny black beetle feeds on the upper and lower leaf surfaces but does not bite through the leaf. In addition to feeding on the plant, the beetle transmits a bacterial wilt disease called Stewart's disease. Sweet corn plants infested in the seedling stage will die or become dwarfed, producing no harvestable ears. Plants affected less drastically may produce normal ears if they receive plenty of water.

Cucumber Beetles

This pest concentrates on vine crops like cucumbers, melons, squash, and pumpkins. The spotted cucumber beetle also likes asparagus, corn, and eggplant. Both major varieties of cucumber beetle (Color Figure 64) —the spotted and the striped— are chewing insects that can cause serious damage and even destroy your entire crop.

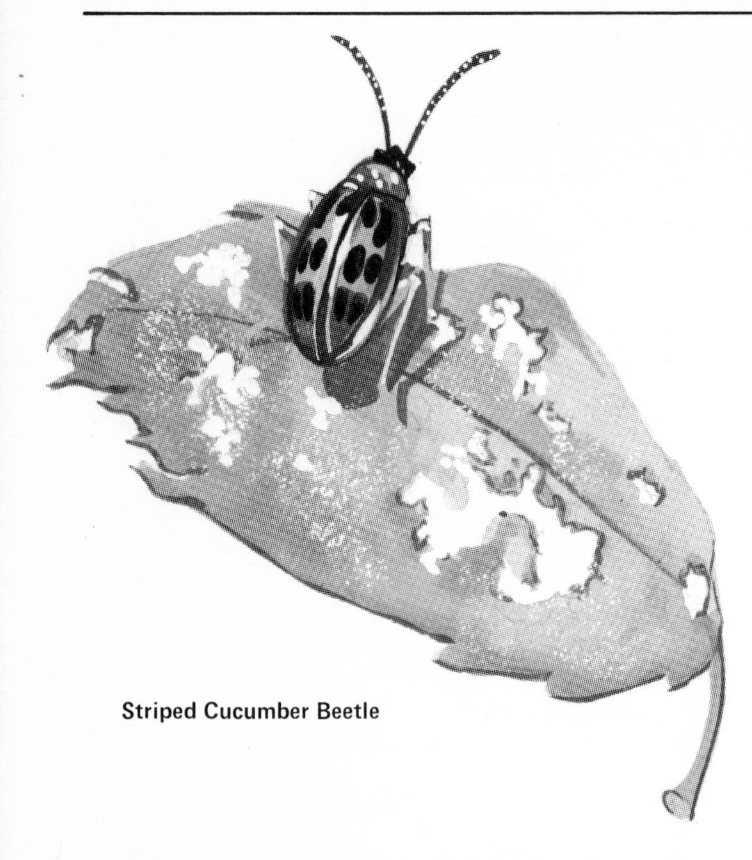

Striped Cucumber Beetle

Just to add to your tribulations, they also spread bacterial wilt and mosaic diseases.

Cutworms

There are three major cutworm species apt to attack your garden—the black cutworm, the armyworm, and the variegated cutworm. They attack all sorts of plants, but are especially fond of corn. They feed on roots and underground stems, eat the leaves, and even cut through the stems near ground level.

Controlling these menaces isn't hard. You can put paper, aluminum foil, or metal collars around the stems of the plants, or apply an insecticide at the first sign of cutting.

Earworms

The earworm is probably the major corn pest throughout the United States. It feeds on the whorl, developing tassel, and ears. It forms a disgusting mold that makes the ears too repulsive to eat. The only effective method of control is the use of insecticides.

Grasshoppers

These devastating insects (Color Figure 68) were among the plagues suffered by the the Egyptians in Old Testament days. They eat almost any garden crop, chewing their way from the outer edges of the leaves into the interior. Even light infestations can cause tremendous damage to your crops.

Hornworms

Hornworms are large green insects marked with white stripes. They are easily identified by a distinctive horn on their rear ends. There are two species likely to be found in gardens—the tomato or northern hornworm and the tobacco or southern hornworm. The adults are moths which do not feed.

Probably the best way of getting rid of them is to pick them off your plants by hand. There is also a parasitic wasp that deposits its eggs under the hornworm's skin (Color Figure 69). The newly hatched wasp larvae then feed on the hornworm, eventually forming small white cocoons on its back.

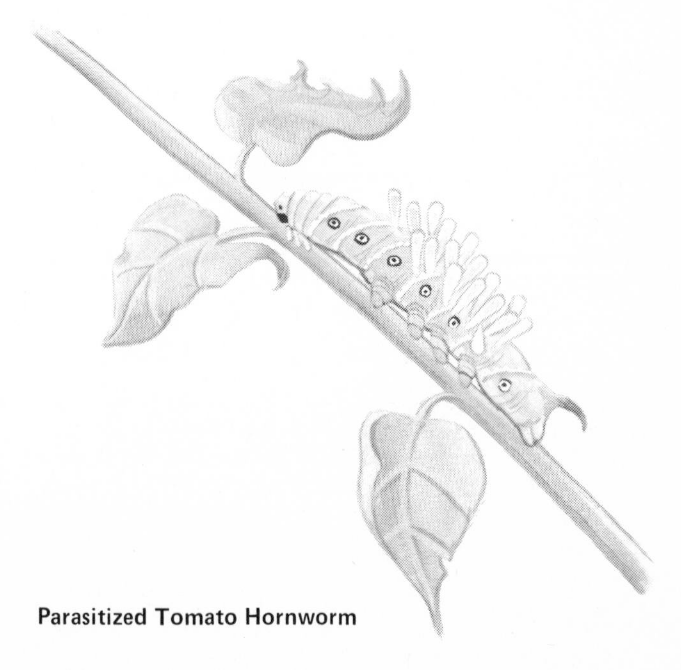

Parasitized Tomato Hornworm

Mealybugs

particularly likely to attack potatoes and beans.

Mealybugs

These oval, soft-bodied creatures are covered with a white powdery substance that gives them a distinctive appearance (Color Figure 76). They're plant-suckers that attack a wide variety of plants. When the mealybug population is high, your plants will look droopy.

Mexican Bean Beetles

This insect is found in every state east of the Mississippi, but it is probably an import from Mexico. It just loves your beans. If you can count to eight, you won't have any trouble identifying this beetle. It's a bright copper-yellow with eight spots on each wing cover.

Iris Borers

As summer progresses, you may notice that your irises are looking ragged. In time, the leaves will turn brown and die and the bases of the plants take on a rotting appearance. (Color Figure 77). The iris borer is the most likely culprit. Controls must be applied early when the larva borer tunnels are first noticed.

Lace Bugs

These plant-suckers are found on the lower surfaces of the leaves. Lace bugs are delicate, oval insects with transparent wings. Look for tiny brown spots on the underside of the leaves.

Leafhoppers

Leafhoppers, discussed in more detail in earlier sections of the book, damage gardens in three different ways. They suck plant juices, turning the leaves brown, inflicting "hopper-burn" (a severe injury caused by a toxic reaction), and also carry diseases from plant to plant. They are

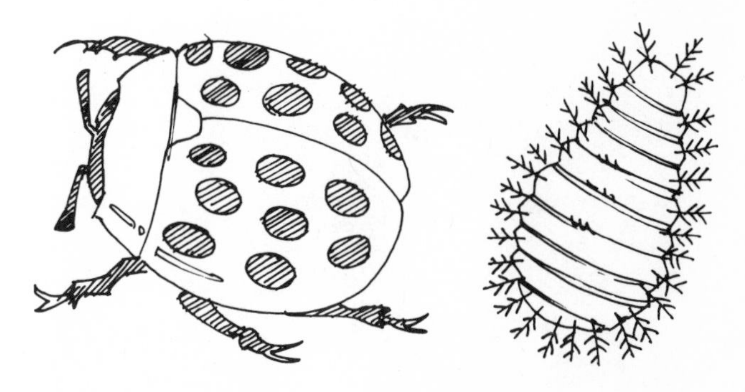

Mexican Bean Beetle Mexican Bean Beetle Larva

Mites

All kinds of foliage appeal to these tiny creatures. Mites rasp and pierce plant tissue, removing juices and chlorophyll. Even though they are so small, a heavy population is easy to detect. Look for tiny specks scurrying around the lower surface of the leaves. If you suspect mites are present but can't see them, strike a few leaves sharply with one hand while holding a piece of white paper under them with the other hand. They'll fall down on the paper and the white surface will make them simple to see.

Picnic Beetles

You may spot these beetles when you're grilling hamburgers outdoors. The smell of cooking food attracts them. They seldom bother healthy plants, but they do enjoy decaying organic material like bruised or overripe fruit. Tomatoes and ears of corn are their favorite dishes. You shouldn't have any trouble identifying them. Look for small black beetles with yellow-orange dots on their backs (Color Figure 67).

Chemical controls usually aren't necessary. Harvest your crops before they become overripe and clean up all rotting plant material quickly.

Rose Scale

Scales

Scale insects are usually not discovered until after the infested plants die. They are small, rather drab-colored creatures that come in a variety of shapes (Color Figure 74). It is sometimes difficult to stop them because they are usually stationary and appear to be part of the plant itself.

The removal of sap by a large population of scales causes the damage. Be particularly alert to this danger.

Slugs and Snails

These pests can be extremely troublesome because they are so difficult to control. Slugs and snails are related to oysters and clams. Look for them early in the spring, especially when dense foliage is close to the ground.

A number of remedies have been suggested as means of controlling them. One gardener in Connecticut placed shallow containers of beer around her garden and found that many slugs crawled into the liquid and drowned. Others have experimented with combinations of bran, wine, vinegar, grape juice, corn cobs, and other nonchemical ingredients.

The standard slug bait pellets available to the home gardener are made of metaldehyde. The fact that this chemical is toxic to slugs and snails was accidentally discovered by a woman in Africa.

Recent controlled experiments revealed that beer and fresh, unfermented grape juice were attractive to slugs and snails. Wine, vinegar, and ethyl alcohol were ineffective. Soaking bait pellets in beer may make them more palatable to these pests.

Slugs and snails like to congregate in sheltered areas, so place the pellets underneath small wooden boards in the areas that are infested. Remember that these pellets are also dangerous to small children and pets. Follow the label directions carefully.

Soil Insects (see, "Lawn Pests")

Squash Vine Borers

If your vine plants look fine one day and then suddenly start wilting, suspect vine borers (Color Figure 65). The larvae attack squash, pumpkin, gourds, cucumber, and muskmelon plants with equal enthusiasm. Once the plants begin dying, there is little you can do. They will not respond to the most encouraging of pep talks.

Prevention is the only cure in this case. Spraying every five to seven days after vining begins will protect your crops from this invisible menace.

Springtails

There are close to 2000 species of springtails. Some can be found on beaches, in water, and even on snow banks. The ones you are most likely to see live on decaying organic matter. Look for very tiny insects that move by hooking and releasing their tails like springs.

Control is simple. Clean out all rotting organic material. Good sanitation usually makes chemicals unnecessary.

Thrips

Thrips rasp and puncture leaf surfaces, feeding on juices and bits of foliage. The first damage shows up as white blotches on leaves or brown streaks on cabbages. Thrips are general feeders and love a good salad of onion, tomato, beets, beans, and flowers. They even like weeds. Look for tiny (only 1/25-inch long) insects from early summer to fall.

White Flies

These tiny white insects are leaf-suckers (Color Figure 75). They excrete "honeydew" which covers the leaf surface, providing a sticky growth medium for a black sooty mold. The mold can interfere with photosynthesis, stunting growth and perhaps causing inferior crops. White flies congregate in large numbers. When a heavily infested plant is shaken, they take to the air, surrounding the plant with a white cloud of flies.

Unfortunately, there is no really effective means of controlling these pests. A promising new insecticide that appears to do a good job has not yet been approved for use on edible plants.

Wireworms

There are many varieties of wireworms, looking very much alike and attacking the roots of different crops. The larvae (Color Figure 66) live in the soil for two to six years before maturing into click beetles. In the larval stage, they drill into tubers or the underground portions of stalks. Occasionally, they destroy seeds before they can germinate. Look for stunting or wilting of the plant. Growth may be distorted and no harvestable crop produced.

Organic Gardening

Many people today are enthusiastic about organic gardening—that is, raising flowers and crops without using chemical pesticides. They maintain that "natural" foods are better-tasting and healthier than those grown with commercial pesticides.

It is important to separate fact from fallacy when discussing this emotional issue. There is no doubt that the agricultural revolution caused by chemical pesticides has resulted in dramatically higher food production. In addition, the efficient elimination of pests has substantially reduced the number of people killed by disease-carrying pests. Although the questions raised by concerned environmentalists are legitimate, perhaps the solution lies in the development of better and safer pesticides, coupled with more discriminating application, rather than in an outright ban against chemicals.

The average home gardener, of course, is not making a significant contribution to the world's food supply, but is reducing the family grocery bill while having fun. The larger questions raised by agricultural mass production are irrelevant. The home gardener who is concerned about the excessive use of chemicals can grow pesticide-free crops in the backyard.

Successful organic gardening, however, does require a serious commitment on your part. Unless your crop is very small and you have a great deal of time and energy to expend on your plants, you cannot hope for good results. Your crops will be smaller and much less attractive than the foods you can buy at the market. Of course, a head of lettuce considered unsuitable for commercial sale because it is stunted or blemished is still edible. If your family is not overly fussy about the appearance of food, you can use these "commercial rejects" to prepare delicious and wholesome meals.

The Scientific Method

Because there are so many recommended organic techniques being publicized, it is vital to know which ones really work and which are merely old wives' tales that may do more harm than good.

Well-conducted research is the cornerstone of the scientific method. Although the research itself may be complex, the method used to conduct it is simple and logical. In essence, it consists of defining the problem, establishing control comparisons, and obtaining reproducible results.

Defining the Problem—What does the experimenter hope to prove? One test is usually able to answer only one question. There must be one unknown rather than several. In the studies described below, the experimenters wanted to learn which commonly recommended organic methods of pesticide control were effective.

Establishing Control Comparisons—In order to rule out the effects of variables such as soil, rain, general growing conditions, etc., the experimenters planted cabbages and other crops that were not treated with any of the controls being tested. Sometimes controls cannot be used; in medicine, for example, witholding needed medicine for control studies would be unethical. In gardening, however, no such ethical questions arise.

Obtaining Reproducible Results—A successful experiment can be duplicated by other scientists or by the same scientists at a later time. A one-time result may or may not be valid. A home gardener, for example, may truthfully say that a mixture of garlic bulbs, mineral oil, soap, and water seems to control cutworms in his garden. A scientist, however, would ask whether cutworms were found in nearby gardens, whether he had raised untreated control crops, whether the mixture worked on other crops, if his crop was one attractive to cutworms in the first place, and whether his results were a fluke or whether they would work again. Until these questions were answered, the mixture could not be considered truly effective.

The University of Illinois Studies—In an effort to determine the validity of some recommended organic treatments, scientists at the University of Illinois conducted two studies.

The first study tested the effect of companion planting. Many organic gardeners maintain that planting marigolds and/or various herbs around the crops discourages pests. The test crops chosen were cabbage, eggplants, and snap beans and the companion plants were hyssop, summer savory, thyme, sage, peppermint, catnip, tansy, marigold, petunia, nasturtium, and onion.

The results clearly showed that the companion plants had no apparent beneficial effect. The crops "protected" by them did not differ from control crops left unprotected.

The second study was designed to test the effectiveness of recommended organic sprays and dusts. The chart shows the sprays and dusts tested and the final results. *Bacillus thuringiensis,* a living disease organism, was the only one that produced good control. Some of the recommended treatments actually produced worse results than no treatment at all.

Some Helpful Techniques

You may be tempted to throw up your hands in despair. Is there anything that works? The answer is yes—if you are willing to tolerate some amount of insect damage and if you are willing to settle for a small crop.

Preparing the Garden—Avoid planting your garden in ground that has been in grass for years. Soil insects such as grubs and wireworms are difficult to eliminate without pesticides. If possible, rotate your vegetable and flower plots each year.

Keep garden areas as free from weeds and debris as you can. Many insects spend the winter in weeds and move to vegetables in the spring. This is particularly true of aphids and stalk borers.

Planting Your Crops—Select hardy crops that are resistant to damage and disease. You are much more likely to be successful with beans, peas, and spinach than with more susceptible crops such as radishes, broccoli, cucumbers and melons.

Interplant whenever possible. Instead of concentrating each crop in neat rows, mix the crops together. If nasty insects devour one plant they'll have a harder time finding the next one.

It's a good idea to use partly grown vegetable transplants rather than seeds. A healthy transplant is better able to resist damage than a tender shoot. You also

RESULTS OF ORGANIC CONTROL EXPERIMENTS

CONTROL	RESULTS			
	Average Weight of Cabbage Heads (pounds)*	Total No. of Heads	Edible Heads	
			No. †	% of Total
Aluminum foil	4.3	12	3	25
Bacillus thuringiensis††	4.4	12	11	91
Carbaryl††	5.1	12	8	66
Garlic and hot pepper	3.0	12	6	50
Lemon juice	2.8	12	7	57
Lux soap	0.0	12	0	0
Onion (shallot)	3.1	12	3	25
Powdered lime	3.9	12	2	16
Rye flour	2.5	12	7	57
Salt	2.2	12	3	25
Salt and flour	2.8	12	4	33
Screens	1.3	12	2	16
Sour milk	1.8	12	3	25
Untreated controls	3.0	12	3	25

* *Note that of the 13 experimental controls, seven produced heads weighing, on the average, less than the untreated heads, one produced heads of equal weight, and five produced heads weighing more.*

† *Note that of the 13 experimental controls, three produced fewer edible heads than the untreated group, four produced the same number, and six produced more.*

†† *While carbaryl, a carbamate pesticide, produced heads weighing the most,* Bacillus thuringiensis, *a living disease organism, produced the most edible heads.*

eliminate the problem of seed damage, which often results in crop failure.

Cultivation—Proper fertilization and watering are essential. Healthy plants are much more likely to withstand insect attacks. Be careful when applying manure and compost. Many insects love decaying organic materials. You don't want to import them into your garden.

Physical Controls—Methods such as handpicking can be effective. If your garden is small and you have the time to spare, you can physically remove insects from each individual plant. This is an excellent means of controlling larger insects, but ineffective against eggs and the many insects too small for you to see. In addition, handpicking must be done at least twice a day to work well.

Mechanical Controls—Equipment such as screens can be helpful but they also have disadvantages. The same screen that keeps insects away from your plants also traps insects inside the garden. If you have insect problems before the screen is set up, these trapped insects will feed—and breed—on your crops, producing substantial damage. Good screens, however, will protect plants. Placing foil collars around plants will protect them against some insects.

Biological Controls—Biological control methods are relied upon by many organic gardeners. Insect pests usually have natural enemies that you can use to reduce their numbers. Establishing an ecological balance between predators and their victims is, however, far from easy.

Beneficial predators do not sign a labor contract. The moment their food supply is exhausted, they move on to greener pastures. Some efficient predators may do a terrific job in one area of the country where their natural prey is abundant but be virtually useless elsewhere where the insect population is different. Some of the following predators may be helpful.

Braconid wasps prey on aphids. Some of these tiny wasps attack only one aphid species while others prey on as many as 30 different aphids. One female wasp can kill hundreds of aphids by inserting one egg into each aphid she encounters. Two or three days later, the egg hatches and the wasp larva begins feeding on the aphid. The larva eventually binds the aphid to a leaf with silk and spins a cocoon inside or beneath the aphid. At the end of the pupal stage, the new wasp cuts a hole through the dead aphid and emerges.

Dragonflies should be among your favorites. One dragonfly eats ten times its own weight in insects each day. The immature or naiad stage is spent in water where the naiads earn their welcome by devouring mosquito larvae.

Fireflies or lightning bugs are really beetles. These interesting creatures not only add a sparkling touch to your garden but also prey on cutworms, snails, and slugs.

Ground beetles shouldn't be confused with the horrid June beetles that do so much damage. Both are large and dark-colored, but the friendly ground beetles should be encouraged because they feed on slugs and caterpillars. They can become pests in the fall when cooler weather makes them look for shelter inside the house.

Lacewings are delicate little creatures that you can recognize by their pale green wings and large eyes. Most adults feed on plant nectars and the "honeydew" secreted by aphids, but some species eat the aphids themselves. It is lacewing larvae who are most helpful to the gardener. The young are often called aphid lions because one larva can consume hundreds of aphids before spinning its cocoon. They also like mealybugs, lice, and thrips.

Lady beetles have received a lot of attention in seed catalogues and organic gardening magazines. You can purchase them by the gallon (one gallon contains 72,000 to 80,000 adult beetles). These commercial beetles are usually collected in mountainous areas where they congregate in enormous numbers, returning to the same spots year after year.

Both larval and adult lady beetles feed on pests. Although aphids form their favorite diet, they also feed on scale insects, mealybugs, and spider mites. One larva can eat 400 aphids during its development. An adult will consume 300 aphids before laying eggs and three to ten for each egg produced thereafter (more than 5000 aphids in its lifetime).

If these commercially available beetles lived up to expectations, they would be invaluable. Studies have shown, however, that they are less useful than you might hope. When the beetles are released in the spring, they disperse quickly and widely. Only a few, therefore, remain in the area they are supposed to protect. Beetles released in the summer, on the other hand, stay in the area but feed and reproduce with abnormal slowness.

"Shipped in" beetles, therefore, are unlikely to repay your investment. The native beetle population in your vicinity will do a much better job of controlling pests than commercial imports.

Praying mantises are sometimes seen as the answer to a gardener's prayers. In fact, however, they won't do you much good. These large, slow-moving insects are striking because of their elongated and powerful front legs, which they use to capture their prey. Since, like lady beetles, they recognize no boundary lines, they will leave as soon as their food supply dwindles. In addition, they are just as likely to eat beneficial insects as harmful ones.

Spiders may travel along the ground (Color Figure 34) or lurk in the hearts of flowers waiting for unwary insects. Still others, perhaps the most familiar, spin sticky webs that capture flies. Although spiders can be useful in eliminating pests, they also serve as food for other creatures. The mud-dauber wasp, for example, paralyzes a spider with a sting and seals it inside the mud chamber where its eggs are deposited. The newly hatched larvae find a

Wolf Spider

tasty treat waiting for them. (It's hard to imagine a spider being tasty, but the larvae just love snacking on it!)

Syrphid flies prey on aphids. The green, legless larvae are voracious eaters, often consuming hundreds of aphids apiece. In fact, a larva can eat one aphid per minute as long as the supply lasts. Syrphid flies are often mistaken for honeybees and are actually good pollinators.

An Effective Natural Control—Perhaps the most efficient organic control is *Bacillus thuringiensis.* This is a disease organism that is very helpful in dealing with certain pests. Its effectiveness varies with the pest species, so it may or may not work in your garden. It is excellent for controlling cabbage worms and the larvae of most moths and butterflies, but is ineffective against beetles and other groups of insects. Because *Bacillus thuringiensis* is a living organism and not a synthetic compound, it is one product that organic gardeners can use without sacrificing their principles.

Conclusions

If you are enthusiastic enough about organic gardening to spend the extra time and energy it requires, you can expect some measure of success. The decision is yours.

Vegetable and Flower Pests

SUGGESTIONS FOR CONTROLLING VEGETABLE PESTS

Pest	Crops	Pesticides	Comments
Aphids	Most crops	Diazinon, malathion	Apply spray directly to foliage beginning early in the season
Bean leaf beetles	Beans	Carbaryl	Spray when insects or damage are seen
Blister beetles	Most crops	Carbaryl	Spray when insects or damage are seen
Cabbage worms	Cabbage, leafy vegetables, and salad crops	*Bacillus thuringiensis*	Spray every seven to ten days when worms are small
Colorado potato beetles	Eggplants, potatoes, tomatoes	Carbaryl	Spray when insects or damage are seen
Corn borers	Sweet corn	Carbaryl	Apply spray every three days to whorl and ear zones when borers are feeding; use at least three applications
Cutworms	Most crops	Carbaryl	Apply spray to base of plants at first sign of cutting
Cucumber beetles	Vine crops	Carbaryl	Spray in the evening hours (to avoid killing bees) when beetles appear in spring at the time of blossoming
Earworms	Sweet corn, tomatoes	Carbaryl	Spray late-maturing tomatoes three or four times at seven-day intervals from the time fruit is small; spray corn every two days (at least four or five times) at fresh-silk stage
Flea beetles	Most crops	Carbaryl	Spray when beetles are seen
Grasshoppers	Most crops	Carbaryl	Spray garden borders when insects are small
Grubs	Most crops	Chlorpyrifos, Diazinon	Drench soil prior to planting
Hornworms	Tomatoes	*Bacillus thuringiensis*, Carbaryl	Spraying is seldom necessary; use hand-picking
Leafhoppers	Most crops	Carbaryl, Diazinon, malathion	Spray when insects are seen or when hopper burn symptoms appear
Mexican bean beetles	Beans	Carbaryl	Apply spray to the undersides of leaves if insects or damage are seen
Mites	Most crops	Malathion	Spray (especially undersides of leaves) when mites are seen
Potato leafhopper	Beans, potatoes	Carbaryl, malathion	Spray at weekly intervals
Picnic beetles	Most crops	Carbaryl	Spray when insects or damage are seen

Vegetable and Flower Pests

Pest	Crops	Pesticides	Comments
Soil insects	All crops	Diazinon	Rake thoroughly into soil prior to planting
Squash vine borers	Vine crops	Carbaryl	Apply spray to crowns and runners weekly when vining begins; apply late in the day to avoid killing bees
Thrips	Most crops	Diazinon, malathion	Spray when damage is seen
Wireworms	Most crops	Diazinon	Drench soil prior to planting

SUGGESTIONS FOR CONTROLLING FLOWER PESTS

Pest	Pesticides[a]	Comments
Ants	Diazinon[b]	Drench soil prior to planting
Aphids	Malathion[c]	Spray foliage generously; repeat if insects or damage are seen
Blister beetles	Carbaryl[d]	Spray foliage; repeat if insects or damage are seen
Cutworms	Diazinon[b]	Spray soil but do not spray blooms; protective collars may suffice for a small number of plants
Grasshoppers	Carbaryl[d], malathion[c]	Spray plants and adjacent grassy or infested areas
Grubs	Diazinon[b]	Drench soil prior to planting
Iris borers	Dimethoate[e]	Spray when insects or damage are seen but do not spray blooms; one application should be sufficient; add detergent (2 teaspoons per gallon of spray) to improve coverage
Lace bugs	Malathion[c]	Spray foliage; repeat if insects or damage are seen
Leaf-feeding beetles	Carbaryl[d]	Spray if insects or damage are seen
Leafhoppers	Carbaryl[d]	Spray if insects or damage are seen
Mealybugs	Malathion[c]	Spray foliage; repeat if insects or damage are seen
Scales	Malathion[c]	Spray foliage; repeat if insects or damage are seen
Slugs	Metaldehyde	Clean up debris and organic materials before spreading pellets; cover pellets with board
Soil insects	Diazinon[b]	Drench soil prior to planting
Spider mites	Dicofol	Spray liberally, paying special attention to the undersides of leaves; repeat if insects or damage are seen
Springtails	Malathion[c]	Apply spray or dust to base and foliage of plants
Stalk borers	Carbaryl[d]	Spray if insects or damage are seen
Thrips	Carbaryl[d]	Spray if insects or damage are seen
Wasps (soil-nesting)	Diazinon[b]	Drench soil prior to planting
White flies	Pyrethrin, resmethrin	Spray if insects or damage are seen
Wireworms	Diazinon[b]	Drench soil prior to planting

[a] *Do not spray plants when they are in full bloom or bees will be killed. Never use oil-base pesticides on foliage because they can cause leaf burn.*

[b] *Do not use Diazinon on ferns.*

[c] *Do not use malathion on African violets*

[d] *Do not use carbaryl on Boston ivy. Repeated use may encourage mite or aphid infestations.*

[e] *Do not use dimethoate on chrysanthemums.*

Small Animal Pests

Cute as they are, these small creatures can invade your home and wreck your garden.

AT ONE TIME or another, every homeowner is faced with unwelcome guests—some of them nonhuman. As suburban areas grow, wild animals find their territories being taken over by large two-legged creatures who chop down trees, divert water supplies, and make a lot of noise. The animals must perish, move, or adapt to the changed environment.

The small animals are often reluctant to move, perhaps because they feel that they were there first. After all, it was their ancestors who pioneered the area. Or perhaps a house just seems like an odd new kind of tree, one full of nooks and crannies that any house-hunting little creature would love. In any case, they end up competing with man and his domesticated animals for the property.

The average person is seldom aware of these wild creatures living so close by, but in fact they are surprisingly numerous. The small animal population is probably ten times that of the more visible—and more audible—birds. In wooded areas, there may be 10 to 60 per acre.

Naturally, these little animals are attracted to a house that offers warmth and shelter in the winter, protection against larger predators, and easily obtained food. That's why you'll find raccoons climbing down your chimney or rooting through garbage cans, squirrels setting up housekeeping in the attic, bats hanging happily from the beams, moles tunneling under the lawn, rabbits munching in the garden, and snakes sunning themselves in the patio. You may find yourself with a second career as a zookeeper!

Many states have strict laws that prohibit the killing or trapping of fur-bearing animals. These regulations, which are intended to protect the creatures from extinction, can create a real headache for the homeowner. If you are one of those rare people who love and like to live with all sorts of animals, more power to you. Most of us, however, are anxious to get rid of these nuisances. Once you discover the problem, prompt action is essential, and obtaining the necessary legal permission to destroy them can be time-consuming.

We cannot recommend using traps or poisons in areas where small children or pets can be found. If the situation is bad enough to warrant such extreme methods, you should call in a professional exterminator who will solve the problem safely as well as effectively.

There are other, less drastic control measures you can take. As you read through these suggestions, remember that not all of them will be practical in your specific situation. They may, nonetheless, give you some ideas that you can use to outwit your troublesome acquaintances.

Bats

Just the thought of a bat in your house is enough to send shivers down your spine! Perhaps it's because they have been associated with witchcraft and black magic. Nonetheless, bats are usually harmless. They don't attack people or get entangled in hair as is commonly supposed. They can, however, carry rabies and you should take them seriously for this reason.

Despite their wings, bats are not birds but flying mammals, the only ones known. They are nocturnal animals, doing all their feeding at night. Because of their mobility, they can feed on the wing, often consuming ten times their own weight in a single night.

Built-in sonar capabilities enable bats to fly quickly and accurately through the darkest night. They send out high-pitched sounds that bounce off objects in their paths, revealing their locations and preventing collisions.

During the daylight hours, bats shelter in caves, hollow trees, attics, and garages, hanging upside down by their rigid curved claws. Most species congregate in groups or colonies. Once a colony settles on your property, you may have trouble. Not only is there an extremely objectionable odor from their urine and droppings (guano), but they are noisy as well. In addition, they may harbor bat bugs, blood-sucking insects much like bedbugs, that can attack sleeping people.

Bat colonies can be difficult to eliminate because they are persistent creatures. Prevention is the better course. The electronic devices now available for discouraging bat infestation are too expensive and too experimental to be recommended. Bat-proofing your property is a good idea in any case, and is an absolute must if you have been infested once and want to eliminate future colonies. The strong odor a colony leaves behind can attract other bats even after the first group has been removed.

Bats enter buildings through unprotected louvers or vents, open or broken windows, and holes in the siding, eaves, or cornices. Some small species can crawl through holes less than one-half inch wide, so you'll have to check carefully.

If you already have bats on the premises, locate their exit and entry routes before closing any holes. Once the first bat leaves in the evening, the rest usually follow within 15 minutes or so. Check the roosting area quickly to make sure they have all departed, and then set to work. Larger openings can be covered with sheet metal or wire mesh and narrow cracks calked. The bats won't return until daybreak, so you should have plenty of time to do a thorough job. If you still find bats inside, you've probably overlooked some holes. Repeat the same procedure until they have all been eliminated.

If you see an occasional bat flying around the house, don't panic. Try to isolate it. A tennis racquet is excellent for knocking it out of the air. Once it is stunned, pick it up with gloves or scoop it up with a piece of paper and deposit it in the trash. Never handle a bat with your bare hands because of the risk of rabies.

Birds

Before taking steps to control bird populations, consult your state or local conservation authorities. Many species are protected by Federal regulations or state laws. Check to see what control measures are allowed and whether a permit is required.

Effective control of avian pests demands proper timing, persistence, and a willingness to experiment. You must take steps as soon as possible after the problem begins. Bird populations are more vulnerable during the nesting period. Control measures must be continued as long as the birds remain. If one form of control doesn't work, try another. Sometimes a combination of techniques is necessary for good results. Noisemakers, although effective, are usually not a good idea in urban or suburban areas. Sticky repellants, pie plates hung from the roof, inflated paper bags, balloons, feather dusters, and imitation snakes have been used successfully.

Blackbirds

The term "blackbird" often includes several species that are considered nuisances. They appear throughout the summer. Grackles, starlings, blackbirds, and cowbirds congregate in large flocks after the mating season. Although many of these birds are protected, you may be allowed to institute controls if they are seriously damaging your garden crops or shade trees. Check with the local authorities.

Pigeons

Because pigeons are fed by bird-lovers and profit from spilled or carelessly dropped food, most cities have large populations of them. In addition, urban buildings provide an abundance of roosting and nesting areas. These large flocks can present a health problem, because pigeons can transmit a number of diseases including encephalitis, histoplasmosis, thrush, and toxoplasmosis.

Pigeon populations can be substantially reduced by removing nests and eggs every two weeks during the spring and summer months. At the same time, potential roosting areas should be eliminated by spreading "glues" or chemical compounds on ledges to make them sticky or by placing "porcupine wire" on them. When using messy glues, apply them on top of removable tape so that they can be stripped off easily when they have served their purpose.

Songbirds

These protected species usually present only minor problems. A woodpecker interested in the cedar on your house, for example, can be controlled by hanging pie plates from the eaves. If a bird nests on a window ledge or near an air conditioner, you may be bothered by bird mites crawling into the house. Remove the nest and use the miticide dicofol to kill the mites.

Sparrows

Destroying nests and eggs at two-week intervals during the spring and summer will take care of this problem. Attach a hook to the end of a long pole so that you can reach up into the eaves and other out-of-the-way spots where nests may be hidden.

Cats

Even domesticated cats can become problems when they stray away from their homes. They soon acquire wild habits, killing birds, rabbits, snakes, and other small creatures. Because they are excellent natural hunters, they can jeopardize the small animal population. Wild cats also become mangy and diseased and sometimes transmit parasites, insects, and illnesses to people and pets. One of their most unpleasant habits is depositing droppings in sandboxes where children play.

Traps that capture these wild cats without harming them are the best solution to the problem. Vagrant cats can be turned over to your local animal control unit and wandering pets can be returned to their owners. Hav-A-Heart traps, which are available at most hardware stores, are efficient and safe. Bait them with fresh or canned fish and set them in areas where the cats hunt or scrounge for food.

Repellents can provide temporary relief. Nicotine sulfate solutions or oil of mustard repel cats without damaging plants. Repeated applications will be necessary, however, because their effectiveness does not last long.

Chipmunks

These cute little creatures resemble ground squirrels but are smaller. Two major groups are found in the United States—the Eastern chipmunk that ranges west to Michigan and south to Florida and the Western chipmunk that is particularly common in the Rockies.

Chipmunks live in underground burrows that sometimes reach 30 feet in length. The entrances may be hard to spot because they are concealed with stones and twigs and have no dirt mound around them to attract attention. Although chipmunks are basically ground-living animals, they are excellent climbers and can scamper up trees to search for food or escape from danger.

They eat insects, nuts, seeds, and berries, but are very seldom a serious threat to crops. If they are damaging your garden, try building them out. Be sure the fence extends far enough underground (six inches to one foot) to discourage burrowing. If there are no small children or pets in the area, you can buy traps at the hardware store and bait them with nuts, pumpkin or sunflower seeds, peanut butter, or corn. You may prefer calling in a professional exterminator.

Gophers

Gophers tend to be underground hermits, living solitary lives except during the brief mating season. Their tunnels may be six inches to several feet deep, and the entrances are easy to spot because of the rounded dirt mounds at the sides.

They feed on underground roots and tubers and may damage these crops. On the other hand, gophers aerate the soil and bury organic matter. If they are doing extensive damage, call in a professional.

Mice (see, Rats and Mice)

Moles

Moles are small burrowing mammals with broad front feet that are designed for shoveling. Although they are sometimes considered pests, they probably earn their keep by eating damage-causing soil insects. Their presence is easy to detect because their burrows leave long, raised ridges along the ground. Eliminating the insects they feed on is usually sufficient to control them (see, "Lawn Pests").

Muskrats

Unless you live close to a lake or pond, muskrats are not going to bother you. These small creatures, much prized for their fur, are semi-aquatic. Their burrows usually have an underwater entrance as well as one just above the water line. Some build little "huts" out of water plants. Muskrats don't construct dams, but their tunneling can weaken banks and contribute to earthfalls.

Probably the best way to prevent burrowing is adding sand or pea gravel to soil. These materials should be placed three feet above and below the water line and should reach a depth of at least one foot. Because the sand and gravel cave in readily, muskrats are discouraged and don't build dens in the area.

Asbestos cement board can also be used as a barrier. The boards are 4 x 8-foot sheets. Be sure the sheet extends two feet above and below the water line. If you need more than one sheet, join the edges smoothly to prevent gnawing.

Opossums

These creatures look like large whitish rats, and are one of the few American species that have pouches for carrying the young. They are nocturnal and live primarily in trees. Because they attack poultry and eat corn, they are considered pests. Hunters value them for their fur and as food. A number of larger animals also prey on opossums. The opossum often misleads these predators by playing dead.

Young opossums are very tiny (about the size of bumblebees) and almost completely transparent. Shortly after birth, they crawl into the mother's abdominal pouch and attach themselves to the mammary glands.

Chemical repellents are usually ineffective. Your best defense is to build them out.

Rabbit

Rabbits

Despite the legend of the Easter Bunny, rabbits are often serious pests. They munch on a wide variety of crops and are extremely fertile (Color Figure 81). Large populations can strip an area of vegetation. In this country, rabbits have so many natural enemies that the average homeowner won't be greatly bothered. In addition to badgers, foxes, and hunting birds that prey on them, rabbits are often the targets of hunters who prize their meat and their fur.

The first thing to do if you have a rabbit problem is to clean out old woodpiles and tall grass and weeds that offer cover and nesting sites.

You can protect trees by wrapping their bases in burlap, tarpaper, or even old newspaper. Repellent chemicals can be used on ornamental plants, but should not be spread on anything edible. These repellents have an odor and/or taste that rabbits find disagreeable. Their effectiveness, however, is questionable. In a recent experimental study designed to test various often-recommended repellents, the rabbits ate all the vegetables.

Raccoons

Raccoons are common throughout the United States, and adapt surprisingly well to urban and suburban environments (Color Figure 79). They feed on a variety of plant seeds, fruits, nuts, insects, frogs, and eggs. In cities, they scavenge for food in garbage cans. Although they are sometimes a nuisance to the gardener, they are often more helpful than not. The insects they devour do more damage than the raccoons do.

If you decide you would be happier without them, a few simple tricks will usually eliminate the problem. Relocating your garbage cans or keeping them brightly lit will discourage adventurous raccoons.

Because raccoons are walking fleabags, you should take action if one sets up housekeeping in your attic or chimney. Chemical repellents such as oil of mustard are temporarily effective, but the obnoxious smell may bother you as much as it does the raccoon. Your best bet is to let the animal leave and then cover its entrance hole with wire mesh.

Raccoon

Rats and Mice

Rats and mice (Color Figure 78) have been mankind's companions since long ago. Rats in particular have caused famines and epidemics throughout recorded history. The dreaded Black Death of the 14th century, which killed between one-fourth and one-third of the population of Europe, was spread from port to port by rats. Rodents have also carried typhoid, dysentery, and rabies organisms. Contamination by rodents can cause acute food poisoning as well. Even today, the deterioration of many of the world's great cities is intensified by their presence. Hungry rats are vicious, and they have been known to attack infants and even sleeping adults.

Traps and poisons may solve the problem temporarily, but building them out is the only effective long-range answer.

Begin by removing potential shelters and food supplies. Rats and mice like to live near old buildings, sheds, piles of trash, or any unclean area. Keep piles of leaves, wood, newspapers, or trash well away from the house. If you are storing boxes, logs, pipes, or other objects outside, keep them 12 to 18 inches off the ground. Keep areas under wooden steps cleaned out.

Rodent Mouse

Solid concrete or masonry steps are more effective than open steps. Fill in holes around the house with tin cans, bricks, concrete, or sheet metal.

You must be especially careful to keep them out of the house. Check every possible entrance. Keep floor drains closed to prevent them from climbing in through the sewers. Stuff the holes around water pipes with steel wool. Patch all cracks and holes in the walls with plaster mixed with steel wool. Holes larger than one-fourth inch should be covered with sheet metal or wire mesh screens. It's also a good idea to check grocery, wood, and coal deliveries for lurking rodents.

Standard snap traps baited with peanut butter, nuts, bacon, cheese, raisins, or chocolate can be very effective. Proper placement of traps is crucial. Rodents usually travel along the walls. If you can't determine their favorite paths, sprinkle some flour or talcum powder in likely areas and look for tracks. Be sure to check the traps frequently.

Poisons should only be used as a last resort. If you must use one, choose a multiple-dose product (available as premixed baits at hardware stores). It may take four days to two weeks of repeated feeding to kill the rodent, but they are safer for people and pets. In case of accidental ingestion, induce vomiting immediately and call your doctor or Poison Control Center for instructions. Poisons should not be used in any house where small children live.

Skunks

Skunks belong to the weasel family and can be found all over the United States. The striped skunk and the spotted or "civet" skunk are the most common varieties (Color Figure 80). They are poor climbers and obtain most of their food on the ground. Skunks eat insects, rodents, eggs, snakes, and a variety of plants. Because they feed on a number of trouble-

Skunk

some pests, skunks are protected by law in many states.

Nonetheless, you won't want one living too close to the house. Like many other wild creatures, they can be carriers of rabies. In addition, the foul-smelling mist squirted from their scent glands in times of danger causes choking and tearing of the eyes. Most animals learn to avoid skunks and, as a result, they tend to be fearless.

When you find a skunk hole, sprinkle some flour around it. Skunks are nocturnal, so go back in the evening and check the flour for tracks. Once you are sure that the hole is empty, seal it with cement, sheet-metal, or wire. Keep the area brightly lit to discourage the skunk from digging another hole nearby. Eliminating the soil insects (see, "Lawn Pests") that are a large part of the skunk's food supply will also tempt it to move elsewhere.

If you or your pet are unfortunate enough to meet a really frightened skunk and are sprayed, neutralize the scent with a deodorant. Some commercial preparations are effective and vinegar or weak chlorine bleach solutions will also work. Tomato juice, one favorite home remedy, won't help.

Snakes

For some reason, most people dislike snakes and are frightened of them. In fact, however, the vast majority of snakes are beautiful and quite harmless. Most are actually beneficial because they feed on insects and rodents. Of the 116 species found in the United States, only 19 are dangerous. Fifteen of these are rattlers, two are water moccasins, and two are coral snakes. Unless you live in an area where these snakes are common, don't worry about any you see slithering across the yard. Even in areas where poisonous snakes are found, bites are rare. More people, in fact, die from wasp stings than from snakebites.

Getting rid of snakes is easier than you might think. Buying a mongoose is unnecessary. The number of snakes per acre is actually fairly small, and persistent killing or removal will reduce the population quickly.

Removing ground cover is an effective means of control. Clear out trash piles and clumps of weed or tall grass. Flagstones, wood piles, loose foundations, rotting stumps and logs, and old cisterns offer shelter to snakes. Either eliminate these hiding places or comb through them frequently to disturb any snakes lurking among them. Make sure adjacent lots are equally discouraging to snakes.

In most cases, the problem is seasonal. Snakes are most active from mid-March to mid-May, with a brief resurgence in fall. During the summer when the ground is dryer and vegetation thicker, they scatter in search of food and are rarely seen.

Squirrels

These small rodents are a familiar sight in most areas of the country (Color Figure 82). Both ground squirrels and tree squirrels forage for roots, seeds, garden produce, and fruits and nuts. The average homeowner is more likely to be bothered by tree squirrels. Some become bold enough to wait on the porch for a friendly human to put out scraps of food.

If you find their chattering annoying or if they are damaging your plants, a few simple measures will minimize the problem.

Squirrel

Squirrels in the attic can be driven out by repellents and/or bright lights. Once they're outside, close entry holes with wire mesh so they can't return. Keep trees trimmed back so that branches are six feet away from the roof. Isolated trees can be protected with bands of sheet metal, two feet wide, placed around the trunk about six feet off the ground. The sheet metal will make it impossible for squirrels to climb up.

A Final Word

Remember that it is easier to share your property with these small animals than to eliminate them completely. If you can, learn to live with them. A good pair of binoculars will permit you to study them and learn a great deal about nature.

Building them out with fences, screens, and sheet metal is the best method of controlling damage. Avoid poisons and traps whenever possible.

Pesticides: How They Work and How To Use Them

In order to use pesticides safely and effectively, you must understand how they work and the proper methods of applying them. Using pesticides wisely is the key to successful pest control.

PESTICIDES are substances or mixtures intended to destroy or control various kinds of pests. The "cide" ending, derived from the Latin word *caedes,* means kill or slaughter. A pesticide, therefore, is literally a pest-killer.

There are hundreds of pesticides now on the market, including insecticides (used to control insects and closely related animals such as spiders and mites); rodenticides (used to control rats, mice, moles, skunks, and some kinds of fish); fungicides (used to control mildews, molds, and slime); herbicides (used to control weeds); and invertebrate animal poisons (used to control jellyfish, snails, slugs, and barnacles). A pesticide listed in one of these categories may also be effective against other classes of pests. Perhaps the best example of this broad-spectrum action is DDT.

Today, all pesticides must be registered with and approved by the Environmental Protection Agency. The official registration number must be shown on the label.

Developing a Pesticide

The pesticide industry must conform to government regulations in order to market a new product. Today, only one compound in 8400 actually makes it to the marketplace. Extensive laboratory and field tests must be conducted by the manufacturer to determine the compound's toxicity, potential health hazards, and environmental impact (both in its original form and after it breaks down chemically). In other words, does it kill pests without endangering the plants it's supposed to protect and other animal—including human—life? Until this question is answered affirmatively, the product cannot be sold.

Using Pesticides Safely

Your first defense against accidental injury or poisoning is the label on your pesticide container. Make sure you read and understand the label information thoroughly before you use the product.

Reading the Label

Government regulations require that every pesticide label contain specific information in a number of important areas. The most important of these items are:

1. The name and address of the manu-

facturer and the EPA registration number. This number includes the initials of the EPA and two sets of numbers separated by a hyphen. The first set indicates the manufacturer and the second set the specific product.

2. Directions for use. These directions tell you exactly how to prepare and apply the pesticide for best results.

3. Precautionary statements. These are warnings of possible hazards and directions to ensure safe handling. The most toxic substances are marked "DANGER—POISON." The word "WARNING" on the label means that the product is not in the most dangerous class, but that extreme care should be used. "CAUTION" appears on those products that are only slightly toxic when used as directed. Pay particular attention to any instructions to keep children and pets away from treated areas and any directions about protective clothing to be worn when applying the chemical.

4. First aid instructions. This section of the label provides emergency information needed in case of accidental poisoning. Read this section carefully before the emergency occurs. When dealing with any kind of poison, you should be prepared.

5. Storage and disposal information. Some pesticides can be kept in storage for

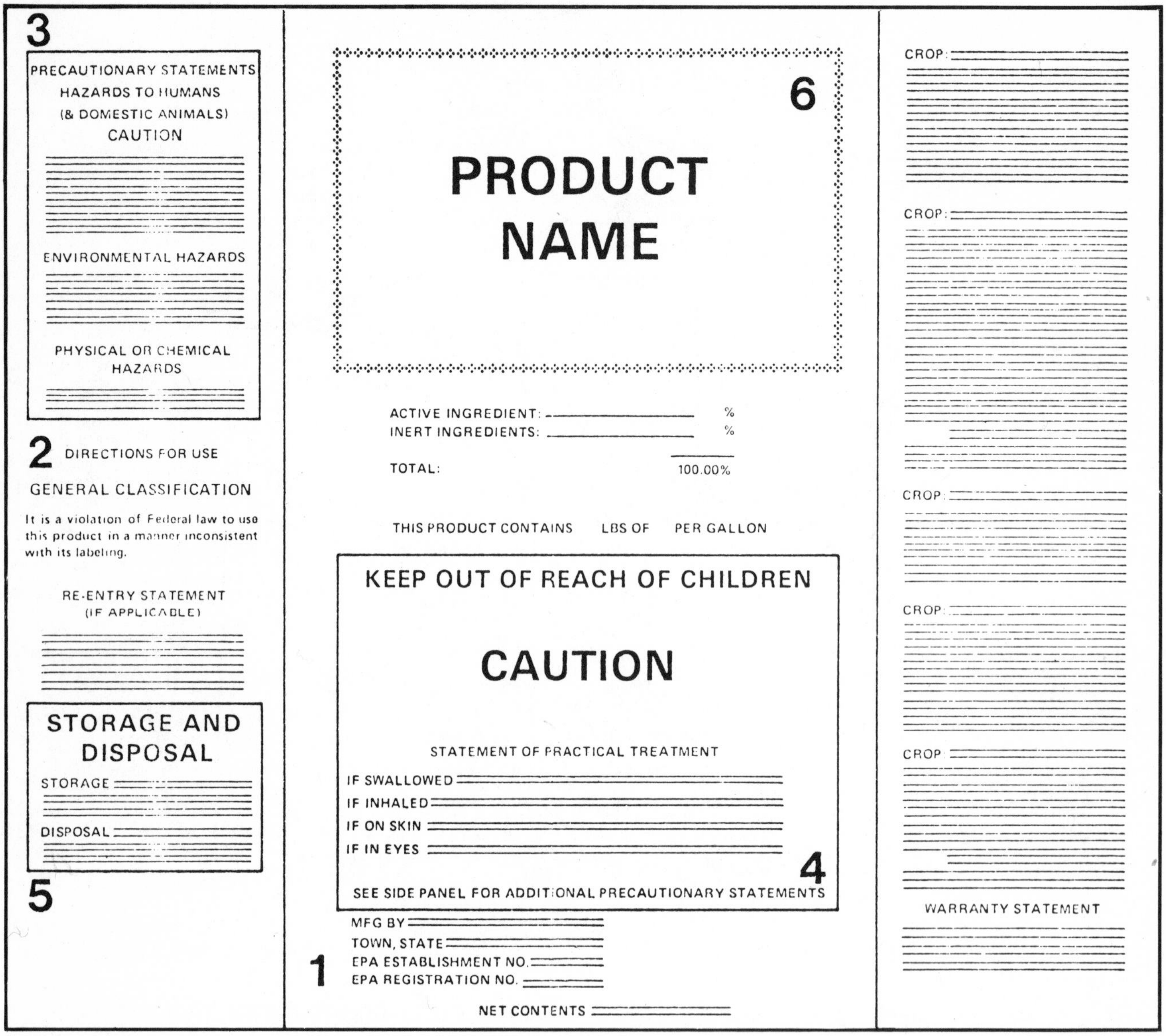

3

PRECAUTIONARY STATEMENTS
HAZARDS TO HUMANS
(& DOMESTIC ANIMALS)
CAUTION

ENVIRONMENTAL HAZARDS

PHYSICAL OR CHEMICAL HAZARDS

2 DIRECTIONS FOR USE

GENERAL CLASSIFICATION

It is a violation of Federal law to use this product in a manner inconsistent with its labeling.

RE-ENTRY STATEMENT
(IF APPLICABLE)

STORAGE AND DISPOSAL
STORAGE
DISPOSAL

5

6

PRODUCT NAME

ACTIVE INGREDIENT: %
INERT INGREDIENTS: %
TOTAL: 100.00%

THIS PRODUCT CONTAINS LBS OF PER GALLON

KEEP OUT OF REACH OF CHILDREN

CAUTION

STATEMENT OF PRACTICAL TREATMENT
IF SWALLOWED
IF INHALED
IF ON SKIN
IF IN EYES

4

SEE SIDE PANEL FOR ADDITIONAL PRECAUTIONARY STATEMENTS

1
MFG BY
TOWN, STATE
EPA ESTABLISHMENT NO.
EPA REGISTRATION NO.

NET CONTENTS

CROP:

CROP:

CROP:

CROP:

CROP:

WARRANTY STATEMENT

later use, but others cannot. They may lose potency or demand special storage conditions you cannot supply. The label will tell you which category your particular product falls into. Disposing of leftover pesticides is not always simple. Some are flammable and others may require some special safety measures. Never dump old containers into the trash without reading the label first.

6. Product name, ingredients, and classification. All pesticides must be classified for general or restricted use. Restricted chemicals are highly toxic and may legally be used only by licensed pesticide applicators. General use means that the product may be sold to and used by the public.

Toxicity and Hazard

Toxicity is the ability of the chemical compound to cause injury. *Toxicity hazard* is the likelihood of this injury occurring. When label directions are followed, the hazard to humans is negligible while the hazard to the target insects is great.

Acute toxicity refers to the effects of a single dose within a short period of time (a few seconds to a few hours). Chronic toxicity refers to the effects of prolonged or repeated exposures over a long period of time.

Lethal Dose 50%

All pesticides (and many other products as well) are given toxicity ratings. They are tested on laboratory animals such as rats, rabbits, or mice. When 50% of a large group of test animals dies, the amount of chemical needed to produce this effect is recorded and known thereafter as the lethal dose 50%, usually abbreviated LD_{50}. This technical term simply means that the dose was lethal for 50% of the test animals.

Acute toxicity ratings are assigned to oral (swallowed) and dermal (absorbed through the skin) doses. Both tests are conducted on the same species of test animal.

The LD_{50} is expressed in terms of milligrams of pesticide per kilogram of the test animal's body weight (mg/kg). The greater the amount of milligrams of pesticide needed to reach the LD_{50}, the less toxic the pesticide is. Thus, an LD_{50} over 5000 mg/kg is considered relatively nontoxic. LD_{50} figures between 501 and 5000 mg/kg are slightly toxic. Figures between 50 and 500 mg/kg are moderately toxic, and any LD_{50} less than 50 mg/kg is very toxic.

Recognizing Poisoning Symptoms

Poisonings resulting from pesticide applications are rare. The majority of poisonings are caused by accidental or deliberate swallowing of the chemical. Just in case, however, you should know what the symptoms of pesticide poisoning are.

The most common symptoms of acute poisoning (single-dose or short-term exposure) are headache, dizziness, nausea, weakness, tightness in the chest area, generalized aching, and possibly muscle twitching. These flu-like symptoms can also be caused by paint, cleaning solvents, and the petroleum derivatives in which most insecticides are dissolved.

Different chemicals may produce distinctive symptoms. There are four basic kinds of insecticides: botanicals, carbamates, organochlorine compounds, and organophosphate compounds.

Botanicals—These are sometimes called natural insecticides because they are made from plants. They are widely used and relatively safe. Reactions to them are allergic rather than a result of toxicity. Symptoms often include stuffy noses, tearing of the eyes, and scratchy throats. People with asthma, hay fever, eczema, or a history of allergic reactions should think twice before using any of these products.

Carbamates—The pesticides in this category are broad-spectrum agents. They cause overstimulation of the involuntary nervous system. Symptoms are salivation, tearing of the eyes, muscle spasms, profuse sweating, nausea and vomiting, diarrhea, disorientation, and even psychotic behavior. Pinpoint pupils and a slow heart beat are other signs helpful in diagnosing carbamate poisoning.

Organochlorine Compounds—The use of these broad-spectrum compounds has been restricted because of their long-lasting residual effects. Use is still legal on some food crops, but their major application today is for termite control. These compounds stimulate the central nervous system and poisoning symptoms include neurological disorders and marked loss of appetite. Large doses cause nervousness and feelings of apprehension, tingling, numbness of the face and extremities, and convulsive movements. These symptoms strongly resemble epileptic seizures.

Organophosphate Compounds—These compounds are used on a wide variety of insects that feed on fruits, vegetables, ornamental plants, and indoor plants. They affect nerve function and may involve danger for the applicator. Use extreme care.

Use With Care—Diagnosing pesticide poisoning is often difficult because the symptoms resemble those of other ailments such as influenza, heat prostration, intoxication, and epilepsy. Make certain your physician is informed that you have been using these products before he diagnoses your illness.

Never use any pesticide unless another adult is at hand. In case of accidental poisoning it is important to have a responsible person to care for the victim. Contaminated clothing should be removed immediately to prevent continued absorption and the skin washed quickly and thoroughly. If the pesticide has entered the eyes, flush them out with large quantities of water. Call a physician immediately.

Storing Pesticides Safely

Once the bug season is over, make sure all pesticides have been destroyed or safely stored. Your clean-up should begin with collecting all pesticide containers. This task will be much simpler if you have kept them together, under lock and key, all summer. Nonetheless, double-check the kitchen, garage, basement, and other areas where they might have been used.

Remember that no pesticide container is ever really "empty." There's always a small amount of chemical left. Read the labels for disposal instructions. As a general rule, bottles should be placed in a sack and broken with a hammer. Cans should be crushed and punctured to prevent reuse. Paper containers should be thrown out immediately after use.

If you are storing pesticides for next year, be sure they are in the original containers. Not only is it important to keep all the label information right at hand, but keeping poisons in tempting pop bottles is asking for tragedy.

Read the label carefully for storage directions. Some pesticides may freeze or break down, losing their effectiveness. Granules and wettable powders must be kept in dry areas well above floor level. Other pesticides are flammable and must be stored in areas where fire is unlikely. Pesticide smoke is highly irritating at best, and may be toxic.

Types of Insecticides

Chemists and entomologists have been trying to develop the ideal insecticide. What characteristics would this perfect insecticide possess?

It would:

1. Be toxic to the insect;
2. Perform its task quickly;
3. Remain toxic long enough to kill the maximum number of insects;
4. Be safe for the user;
5. Be easy to apply;
6. Be compatible with other insecticides and fungicides;
7. Remain in suspension when used as a spray;
8. Be inexpensive.

It would not:

9. Leave dangerous residues;
10. Be absorbed and stored in plant and animal tissues;
11. Break down in storage;
12. Injure the plants and animals it is designed to protect;
13. Have any harmful effect on the soil;
14. Corrode or otherwise damage spray equipment.

No such insecticide now exists, but this ideal is the goal toward which the industry is working. In the meantime, you will find many different forms and mixtures of insecticides at the garden center or hardware store near you. Here's a brief summary of each type and its good and bad points.

Aerosol Sprays

Aerosol sprays almost always contain a mixture of insecticides. They are useful for treating small areas or houseplants (take them outside before spraying!). Aerosol sprays are too expensive for large areas such as gardens.

Dusts

Dusts are made of finely ground talc or clay particles that are mixed with an insecticide. They are sold in a number of different strengths and are easy, even if somewhat messy, to apply. One major advantage offered by dusts is that you don't have to clean an applicator; you simply hang the duster up again.

Dusts seldom harm plants if you use the recommended strengths. In addition, they are helpful in treating narrow cracks and crevices where liquids are difficult to apply. You will need a greater amount, however, because dusts tend to blow away during application and because they contain a smaller percentage of the active insecticide. In addition, they are quite visible. Rain washes them off rather easily, but the dusts will adhere to the foliage better if you apply them early in the morning when the plants are wet with dew.

Granules

These particles are larger than those found in dusts. They are impregnated with an insecticide, an herbicide, or a fertilizer, or some combination of these three. Strengths range from 1% to 33%. Their primary usefulness to the home gardener is in the area of lawn treatment. Apply them directly on the selected area with a fertilizer spreader. Make sure the spreader is properly calibrated.

Liquid Concentrates

As their name implies, these insecticides are sold in concentrated strengths ranging from 18% to 72%. The active ingredient is dissolved in a solvent to which emulsifying agents are added. The concentrate is mixed with water and the insecticide kept in suspension during spraying. Usually, only a small amount of concentrate is needed to make the spray effective.

These sprays are excellent for surface coverage. Because they are nonflammable, you can use them near open fires and heating units. They do conduct electricity, however, so stay away from wall sockets. If you plan to use them indoors, test spray an inconspicuous spot on the wall to see if any staining occurs. They will not penetrate wood as an oil solution does, but some surfaces may be marred by the water.

On the whole, liquid concentrates are probably the best choice for the average homeowner.

Oil Solutions

Oil solutions contain an insecticide mixed with a highly refined oil base. They are primarily used for controlling indoor pests such as cockroaches, silverfish, and cereal insects. The available, ready-to-use strengths range from 0.5% to 5% depending on the insecticide.

These oil solutions flow readily into cracks and crevices and penetrate wood. They kill insects on contact more quickly than other types of insecticides because the oil penetrates the waxy outer layer of the insects' bodies.

Oil concentrates do not conduct electricity, and, therefore, can be used safely near electrical outlets. On the other hand, you must be alert to the possibilities of burning and staining of surfaces and lingering odor. Never use oil solutions on foliage, because they are likely to cause leaf burn.

Wettable Powders

Wettable powders are similar to dusts, but have an extra ingredient that makes them mix well with water. They are sold in strengths ranging from 15% to 80%.

Wettable powders are better for green foliage that may be injured by oil-base insecticides. They do, however, have two disadvantages. Because there is no emulsifier to keep the insecticide in suspension, the powder has a tendency to settle. You have to shake your sprayer often to keep the solution well mixed. In addition, wettable powders can clog sprayer nozzles. Make sure your equipment is thoroughly clean after each use.

The Individual Pesticides

The following summaries describe each of the pesticides mentioned in this book. These are the only pesticides we can recommend for use by the nonprofessional. Make certain you have read and understood the label information before using any pesticide.

Acephate

Acephate is a broad-spectrum systemic insecticide with residual effects lasting 10 to 15 days, It is very effective against aphids, bagworms, cankerworms, and other leaf-feeders. It is safe to handle.

Bacillus thuringiensis

This insecticide differs from the others discussed in that it is microbial rather than chemical. It is excellent for controlling bagworms, mimosa webworms, and caterpillars, but is less effective against other insects. Sunlight quickly destroys its potency, so there is little residual activity.

Bacillus thuringiensis is an excellent choice for the home gardener because it has no toxic effect on warm-blooded animals, including man. In addition, bees and most other beneficial insects and the plants themselves are not affected by it. There is no need to wait between applications and harvest.

Carbaryl

Carbaryl is used in flea and tick powders as well as for controlling fruit and vegetable pests. It is a broad-spectrum pesticide with residual effects that last 10 to 12 days if it doesn't rain. It is relatively safe to handle and may be applied to many fruits and vegetables just before harvesting (see chart for specific items). It is particularly effective against sod webworm larvae and is often used in fertilizer compounds.

Chlorpyrifos

This broad-spectrum insecticide is used to control turf and ornamental plant pests, household insects such as cockroaches, and nuisances such as mosquitoes. Its residual effect lasts about 30 days indoors and 10 to 12 days outside.

Deet

Deet is a repellent to keep mosquitoes, biting fleas, gnats, chiggers, ticks, and other nuisance insects away from people. It is very effective and safe for use on human skin, but does require frequent application.

Diazinon

This multipurpose insecticide is a powerful weapon in the home gardener's arsenal. It is very effective against soil insects and a variety of insects that attack vegetables and fruit trees. You can also use it indoors to control nuisance pests such as cockroaches and cereal bugs.

Under normal conditions, a spray used outdoors will remain effective for seven to ten days. As a soil insecticide, its effects last seven to ten weeks. Rake it thoroughly into the soil just before planting. It remains potent for 30 days or more when used indoors. Do not use Diazinon on ferns or hibiscus.

Dichlorvos

Dichlorvos is a contact and stomach poison that is commonly used in pest strips. When combined with diazinon and chlorpyrifos, it is effective against cockroaches. Although it works very well against flying insects, a dichlorvos spray is effective for only a few hours. The impregnated strips last much longer, retaining their potency for up to three months. Smaller strips hung on the lid of a garbage can are excellent for controlling flies.

Dicofol

This miticide does a good job of controlling a variety of mite species that feed on fruits, vegetables, and ornamental plants. The residual effects last about ten days.

Dimethoate

Dimethoate is very effective for controlling iris borers and does a fair job on borers in general. It is used as a wall spray to eliminate houseflies in doghouses and other buildings that house animals and can be used on ornamental plants of various kinds (with the exception of chrysanthemums). Liquid concentrates and wettable powders are available on a limited basis.

Dormant Oils

Dormant oils derive their name from the fact that they are applied during the winter when plants are dormant. These products are petroleum oils that control scale insects, aphids, and mites. Oil-base sprays should not be applied directly to foliage because they may cause leaf burn.

Lindane

Lindane belongs to the organochlorine group. These compounds have long-lasting residual effects and should not be used indoors. They do an excellent job of controlling powder-post beetles. They are available in various forms, including a smoke compound that cannot be recommended.

Malathion

Malathion is probably the most widely used insecticide in the United States. It's even used as far away as South Africa and the Soviet Union. Malathion effectively controls a large number of pests, including aphids, spider mites, scale insects, houseflies, and mosquitoes.

It has the advantage of being safe for mammals such as pets. Unlike some insecticides, its toxicity is about the same as that of aspirin. It is also biodegradable and breaks down into substances that do no environmental damage.

In general, you should apply it early in the morning or in the evening because mid-day heat will vaporize it. Its effects will last for about three days. It should not be used on African violets or red cedars.

Metaldehyde

Metaldehyde is a slug and snail bait that can be used around ornamental plants. It should not be used in vegetable gardens or near any edible plants. You can purchase metaldehyde pellets at most garden centers.

Methiocarb

This pesticide is used to control many species of insects and mites that attack fruits, vegetables, and ornamental plants. It is also effective against slugs and snails. Because methiocarb belongs to the carbamate group, it is toxic to bees, and you must be careful to avoid blooms when applying it.

Methoxychlor

This pesticide is widely used because it combines a long-lasting residual action against insects with only a small toxic effect on warm-blooded animals. Its residual effect lasts 90 to 120 days, a fact that eliminates the need to repeat spray applications at frequent intervals. Methoxychlor is effective against the elm bark beetle which carries Dutch elm disease and is often found in multipurpose fruit tree sprays.

Penta

This preservative protects wood from fungus decay, termites, and beetle attack. It is usually available as a liquid concentrate.

Propoxur

This useful product is effective against insect pests such as cockroaches and mosquitoes, and certain lawn and turf pests. Indoors, its effect lasts up to 45 days. Propoxur comes in a number of different forms.

Pyrethrin

Pyrethrin is a botanical insecticide derived from the chrysanthemum. It is probably the safest, least toxic insecticide you can use. Because it leaves no harmful residues on food crops, it can be used in food processing and storage areas. The one major disadvantage of pyrethrin is that it loses its effectiveness in three hours or less, so it is useless when longer-lasting residual activity is needed. It is available almost exclusively in aerosols, often in combination with other ingredients to provide flying insect control.

Resmethrin

This insecticide is safe for household use because it is virtually nontoxic. Houseplant fanciers find it useful in controlling white flies. Its residual action is fairly long-lasting. You must remember, however, that resmethrin has not been approved for use on vegetables or other edible plants.

The Pesticide Checklist

Acephate

Chemical name: *O,S*-dimethyl acetylphosphoramidothioate
Trade name: Orthene
Pesticide category: Organophosphate compound
Toxicity rating: Slightly toxic
Signal word on label: Caution
Acute oral LD_{50}: 866 mg/kg
Available as: Aerosol, liquid concentrate

Bacillus thuringiensis

Chemical name: *Bacillus thuringiensis*
Trade names: Biotrol, Dipel, Thuricide
Pesticide category: Microbial product
Toxicity rating: Harmless to humans and warm-blooded animals
Signal word on label: None
Acute oral LD_{50}: Not applicable
Available as: Liquid concentrate, wettable powder

Carbaryl

Chemical name: 1-naphthyl *N*-methylcarbamate
Trade name: Sevin
Pesticide category: Carbamate
Toxicity rating: Slightly toxic
Signal word on label: Caution
Acute oral LD_{50}: 500 mg/kg
Available as: Dust, granules, liquid concentrate, wettable powder

Chlorpyrifos

Chemical name: *O,O*-diethyl *O*-(3,5,6-trichloro-2-pyridyl)-phosphorothioate
Trade names: Dursban, Lorsban
Pesticide category: Organophosphate compound
Toxicity rating: Moderately toxic
Signal word on label: Warning
Acute oral LD_{50}: 163 mg/kg
Available as: Aerosol, granules, liquid concentrate

Deet

Chemical name: N, N-diethyl-*M*-toluamide
Trade names: Kik, Off
Pesticide category: Insect repellent
Toxicity rating: Slightly toxic
Signal word on label: Caution
Acute oral LD_{50}: 2000 mg/kg
Available as: Aerosol, lotion, roll-on, towelettes

Diazinon

Chemical name: *O,O*-diethyl *O*-(2-isopropyl-4-methyl-6-pyrimidinyl) phosphorothioate
Trade names: Diazinon, Spectracide
Pesticide category: Organophosphate compound
Toxicity rating: Moderately toxic
Signal word on label: Warning
Acute oral LD_{50}: 350 mg/kg
Available as: Dust, granules, liquid concentrate, wettable powder

Dichlorvos

Chemical name: 2,2-dichlorovinyl *O,O*-dimethyl phosphate
Trade names: No-Pest, Vapona
Pesticide category: Organophosphate compound
Toxicity rating: Moderately toxic
Signal word on label: Warning
Acute oral LD_{50} : 56 mg/kg
Available as: Aerosol, impregnated strips, liquid concentrate, oil solution, wettable powder

Dicofol

Chemical name: 1,1-bis (chlorphenyl) -2,2,-trichloroethanol
Trade name: Kelthane
Pesticide category: Organochlorine compound
Toxicity rating: Slightly toxic
Signal word on label: Caution
Acute oral LD_{50}: 809 mg/kg
Available as: Liquid concentrate, wettable powder

Dimethoate

Chemical name: *O,O*-dimethyl *S*-(*N*-methylcarbamoylmethyl) phosphorodithioate
Trade names: Cygon, De-Fend
Pesticide category: Organophosphate compound
Toxicity rating: Moderately toxic
Signal word on label: Warning
Acute oral LD_{50}: 320 mg/kg
Available as: Liquid concentrate, wettable powder

Lindane

Chemical name: Gamma isomer of 1,2,3,4,5,6-hexachlorcyclohexane
Trade names: Forlin, Gamaphex
Pesticide category: Organochlorine compound
Toxicity rating: Moderately toxic
Signal word on label: Warning
Acute oral LD_{50}: 88 mg/kg
Available as: Aerosol, dust, granules, liquid concentrate, oil solution, smoke bomb (not recommended), wettable powder

Malathion

Chemical name: *O,O*-dimethyl *S*-(1,2-dicarbethoxyethyl) phosphorodithionate
Trade names: Cythion, Malaspray
Pesticide category: Organophosphate compound
Toxicity rating: Slightly toxic
Signal word on label: Caution
Acute oral LD_{50}: 1375 mg/kg
Available as: Dust, liquid concentrate, wettable powder

Metaldehyde

Chemical name: Polymer of acetaldehyde
Trade names: Antimalace, Halizon, Namekil
Pesticide category: Slug and snail repellent
Toxicity rating: Slightly toxic
Signal word on label: Caution
Acute oral LD_{50}: 630 mg/kg
Available as: Baits

Methiocarb

Chemical name: 3,5-dimethyl-4-(methylthio) phenol methylcarbamate
Trade name: Mesurol
Pesticide category: Carbamate
Toxicity rating: Moderately toxic
Signal word on label: Warning

Acute oral LD_{50}: 87 mg/kg
Available as: Baits, dust, wettable powder

Methoxychlor

Chemical name: 1,1,1-trichloro-2,2-bis (*p*-methoxyphenol) ethanol
Trade name: Maralate
Pesticide category: Organochlorine compound
Toxicity rating: Relatively nontoxic
Signal word on label: Caution
Acute oral LD_{50}: 6000 mg/kg
Available as: Liquid concentrate, wettable powder

Penta

Chemical name: Pentachlorophenol
Trade names: Dowicide, Weedone
Pesticide category: Organochlorine compound
Toxicity rating: Moderately toxic
Signal word on label: Warning
Acute oral LD_{50}: 50 mg/kg
Available as: Liquid concentrate

Propoxur

Chemical name: 2-(1-methylethoxy) phenol methylcarbamate
Trade name: Baygon
Pesticide category: Carbamate
Toxicity rating: Moderately toxic
Signal word on label: Warning
Acute oral LD_{50}: 95 mg/kg
Available as: Aerosol, baits, dust, liquid concentrate, wettable powder

Pyrethrin

Chemical name: Derived from *Chrysanthemum cineraiaefolium*
Trade name: Pyrethrin
Pesticide category: Botanical
Toxicity rating: Slightly toxic
Signal word on label: Caution
Acute oral LD_{50}: 1500 mg/kg
Available as: Aerosol

Resmethrin

Chemical name: (5-benzyl-3-furyl) methyl-2,2-dimethyl-3-(2-methylprophenyl) cyclopropane carboxylate
Trade names: SBP 1382
Pesticide category: Synthetic botanical
Toxicity rating: Slightly toxic
Signal word on label: caution
Acute oral LD_{50}: 4240 mg/kg
Available as: Aerosol

Spraying Equipment

Effective pest control depends not only on the pesticide you use, but also on the type of equipment you employ and the care you take of it. The equipment designed to do one job may be inadequate for another, and the best equipment will not function properly if you are negligent in maintaining it.

Selecting Your Equipment

Your needs depend on the kinds of plantings you have, and the house you live in. An apartment-dweller may need only a small aerosol intended to control indoor insect pests, while the homeowner with trees and fairly extensive grounds may require more elaborate power spraying equipment.

The following list briefly describes the most readily available kinds of equipment. Read through the list to see which types best fit your needs.

Aerosol Bombs—These pressurized spray cans are small individual applicators intended to provide coverage over small areas. They are excellent for temporary control of flies, mosquitoes, and indoor pests. The apartment-dweller who has an occasional insect problem but who doesn't want to invest in elaborate equipment will find them handy and easy to use. Most of the chemicals sold in other forms for large-scale spraying can be purchased in aerosol cans.

Dusters—The duster is convenient for many small jobs. Dusts are easy to apply and can be purchased ready to use. This means that you don't have to worry about mixing or diluting chemicals. In addition, there is no need to clean out the application equipment afterwards. Dusts do have some disadvantages, however. They must be applied frequently and they often cost more than other pesticides.

There are two main types of duster. The *plunger* duster is a small air pump with a cavity to hold the dust. Small units with eight-ounce capacities are available for indoor use. Larger models with capacities up to five pounds are used outside. *Crank dusters* deliver a continuous flow and are designed for more extensive grounds. Capacities range from 5 to 25 pounds.

Sprayers—Compressed air sprayers are the most popular and versatile models available to the home gardener. Pumping increases the air pressure, which forces the spray out as soon as a quick-action valve is opened. These sprayers, which can be carried by hand or mounted on a cart, can be used for almost every pest control job, both indoors and outside. Capacities range from one to six gallons.

Continuous sprayers deliver a coarse spray as long as pressure is maintained. The user forces the liquid out by pumping a handle that builds air pressure in the tank. Some models have adjustable or changeable nozzles for space spraying. Capacities range from 20 ounces to 3 quarts.

Hose-end sprayers are light and handy, but do have limitations. They are good for fertilizing and applying lawn insecticides, but lack the accurate placement needed for other jobs. Because the sprayer attaches directly to your garden hose, be sure it contains a device to keep the spray from seeping back into your water supply.

Intermittent sprayers are similar to continuous sprayers, but have smaller capacities. They are inexpensive, easy to operate, and do a good job of applying the diluted liquid concentrates used to control flies, mosquitoes, roaches, moths, and silverfish.

Knapsack sprayers, as their name suggests, are carried on the operator's back, thus combining a fairly large amount of spray with mobility. A hand-operated hydraulic pump delivers a steady flow. Capacities range from four to six gallons.

Power sprayers may be necessary for the homeowner with extensive grounds or several fruit trees. Power sprayers designed for home use are available with capacities of ten gallons or more. Power equipment is a definite advantage in tackling big jobs.

Cleaning and Storing Your Equipment

If you spray the lawn with an herbicide, rinse out the sprayer with a garden hose, and then spray the shrubbery with an insecticide, you'll probably end up with dead or dying shrubs! *Thorough* cleansing is required after each use. If you can afford it, separate sprayers for different kinds of pesticides will simplify matters. If you don't want to invest that much money in equipment, one sprayer will suffice, as long as you are careful to remove all spray residues.

Spray residues can corrode or rust your equipment, reducing its usefulness substantially. In addition, if the residues belong to a pesticide group incompatible with the second chemical you use, plant damage, increased toxicity, and loss of effectiveness may occur. The many pesticide mixtures now on the market have been carefully tested to ensure compatibility. Compatibility charts are available, but as a general rule, you should avoid creating your own mixtures. Even chemicals that are compatible may not be particularly effective in combination.

If you always follow these procedures, you should have little difficulty.

Cleaning:

1. Empty spray materials after each use. Dispose of excess materials in a safe place out of the way of children and animals.
2. Flush out every part of the equipment with clear water to prevent clogging.
3. Circulate a detergent-water mixture through hoses and nozzles.
4. Drain the equipment thoroughly.
5. Break down nozzles and clean screens.

These steps are sufficient for water-soluble chemicals. If you have used phenoxy herbicides or brushkillers, an ammonia or charcoal rinse must be added between steps 2 and 3.

Ammonia—Fill your equipment one third to one half full with ammonia solution (one ounce of ammonia per gallon of water). Run part of the mixture through the hoses and nozzles. Allow the remaining solution to stand in the equipment overnight, and then run it through the nozzles. Drain, and rinse thoroughly with clear water.

Charcoal—Fill your equipment one third to one half full with the charcoal mixture (one ounce of activated charcoal and one ounce of soap powder per three gallons of water). Agitate vigorously to distribute the charcoal, and run through the hoses and nozzles. Drain, and rinse thoroughly with clear water.

Storage:

1. Remove hoses and store coiled on a shelf or in a large pail. Do not bend or hang them over nails on a wall.
2. Invert the sprayer when storing to permit thorough drainage.

Index

B

Index

Index

J

L

M

P

Index